Pvt. Ltd.

Evolutionary Computations *for* Manufacturing

Padmakar J. Pawar

Professor and Head

Department of Production Engineering

K. K. Wagh Institute of Engineering Education and Research

Nashik, Maharashtra – 422 003

2019

Studium Press (India) Pvt. Ltd.

Evolutionary Computations *for* Manufacturing

ISBN: 978-93-85046-52-0

Published by:

Studium Press (India) Pvt. Ltd.
4735 / 22, 2nd Floor, Prakash Deep Building
(Near Delhi Medical Association),
Ansari Road, Darya Ganj, New Delhi-110 002
Tel.: + 91-11-43240200-15 (15 lines); Fax: 91-11-43240215
E-mail: pubdir@studiumpress.in

Printed at India

About the Author

P. J. Pawar is currently working as a Professor and Head of Production Engineering Department of K. K. Wagh Institute of Engineering Education and Research, Nashik, Maharashtra (India). He has more than 22 years of teaching experience. He obtained his BE degree in Production Engineering from Savitribai Phule Pune University Pune, M. E (Mechanical Engineering) from Shivaji University Kolhapur and Ph. D. in Mechanical Engineering from SVNIT Surat. His research interests include manufacturing optimization, soft computing, industrial automation and tool and die design. To his credit he has more than 80 publications in reputed international journals and conference proceedings. He also handled/handling projects funded by Savitribai Phule Pune University and AICTE. He has also filed three patents. So far he guided 20 PG students and one Ph. D. student. Presently six research scholars are pursuing Ph. D. under his guidance. He also authored two book chapters for an Edited book titled 'Artificial Intelligence in Manufacturing Research' published by Nova Science Publishers Inc. New York, USA. He has organised several training programs for faculty on the themes related to soft computing and artificial intelligence. He had been invited as an expert speaker on the topic 'Manufacturing Optimization' at several faculty development programs organized by Institutes/organisations of national repute such as SVNIT Surat, NIT Agartala, Indian Space Research Organization (ISRO) Ahmadabad, Bhaba Atomic Research Centre (BARC) Mumbai, JNTU Hyderabad, etc. He is member of professional bodies such as Indian Institution of Production Engineers (IIPE), Indian Institution of Industrial Engineering (IIIE), and Institution of Engineering and Technology (IET).

Preface

In today's competitive world, manufacturing is a diverse sector and contributes enormously to the global economy. In the present scenario of Industry 4.0, cognitive computing plays a vital role. Cognitive computing makes the manufacturing system capable of anticipating new problems, modeling possible solutions and makes decisions by its own. Evolutionary computing being subset of cognitive computing, its acquaintance is very essential to explore the applications of technological drivers of Industry 4.0 such as E-manufacturing, machine learning, augmented reality etc. Evolutionary computational methods are mainly based on biological, molecular or neurological phenomenon that mimics the metaphor of natural biological evolution and/or the social behavior of species. This book provides theoretical concepts and practical applications of several successful computational methods such as genetic algorithms, particle swarm optimization, artificial bee colony algorithm, shuffled frog leaping algorithm, simulated annealing, harmony search algorithm, teaching learning based optimization, etc. As these algorithms are 'intelligent', they offer flexibility to tune their parameters to suit to specific application to achieve best possible solution. Several techniques are discussed in this book to tune these algorithm specific parameters. Fuzzy optimization is a new concept and is discussed in this book through demonstration of how fuzzy logic can be well applied to tune algorithm specific parameters of above mentioned methods. Methods for multi-objective optimization both, priori and posteriori are discussed in detail through real life case studies. Multi-objective optimization through posteriori approaches is an emerging area of research. Three such algorithms namely NSGA, NSGA-II, and multi-objective ABC algorithm are also explained.

The theoretical concepts and philosophy of evolutionary algorithms is demonstrated through several real life practical case studies in three broad categories: improved manufacturing systems, process improvement through parametric optimization, and artificial intelligence in manufacturing. All algorithms are well demonstrated through a simple example in a step by step manner followed by applications of these algorithms to the real life industrial problems. Several case studies are discussed to cover wide range of optimization issues related to manufacturing systems and most of them are based on the research work of the author. The variants of the basic evolutionary algorithms are also discussed in detail.

Author places here in record his profound gratitude to Shri. Balasaheb Wagh, Chairman of K.K. Wagh Education Society, Nashik and Dr. K.N. Nandurkar, Principal, K.K. Wagh Institute of Engineering Education and Research, Nashik for their kind co-operation and permission to publish this book. Author is also grateful to Dr. R.V. Rao, Professor, SVNIT Surat for his guidance and motivation to write this book. Equally, the author feels grateful to M/s Studium Press and the officials there for showing their interest to publish this book. The author's sincere thanks are due also to his family members, especially Ketan Pawar and Aditi Pawar, for timely help.

Table of Contents

1

Introduction

In present customer driven market scenario, customers want customized, personalized and unique products over standardized ones and prefer local, smaller producers over large-scale global manufacturers. But at the same time they also expect locally -produced products to be as cheap and reliable as those internationally produced. This has resulted into several challenges the industry needs to face such as improving productivity along with quality, improving reliability and efficiency of the manufacturing systems, making manufacturing machines capable to make their own decisions, optimize and enhance resources such as labor, workforce, energy, etc. To overcome these challenges, industries must assimilate artificial intelligence in manufacturing through automation, machine learning, computer vision, and cognitive methods. Artificial intelligence comprises of several tools to solve the most complex problems in computer science and evolutionary computations is one of them. Evolutionary computations uses a form of optimization search either through evolutionary process (*e.g.,* genetic algorithm) or through distributed search process (*e.g.,* swarm intelligence). Optimization search in case of manufacturing system is usually a difficult work where the following aspects are required: knowledge of manufacturing system, empirical equations to develop realistic constraints, specifications of resource capabilities, development of an effective optimization criterion, and knowledge of mathematical and numerical optimization techniques. Moreover, performance of these systems is also affected by many factors and a single parameter change will influence the system in a complex way. Because of the many variables and the complex and stochastic nature of the process, achieving the optimal performance is a great challenge. An effective way to solve this problem is to discover the relationship between the performance of the process and its controllable input parameters by modeling the process through suitable mathematical techniques and optimization using evolutionary computing techniques.

The first necessary step for optimization of manufacturing system is to understand the principle governing the manufacturing system. The second step is to identify the attributes affecting the performance measures of manufacturing system. The third step is to develop an explicit mathematical model which may be mechanistic and empirical. The modeling techniques of input-output and in-process parameter relationships are mainly based on mathematical modeling, statistical regression, artificial neural network, and fuzzy logic. The final step is to apply appropriate optimization technique. In general, the optimization techniques can be classified in two distinct types: traditional techniques and non-traditional techniques. Fig.1.1 shows few examples of each type. Although several software and program codes are available for these techniques, one must know the mechanism of working of these techniques for following reasons:

1. To identify new areas of application of these techniques.
2. Limitations of existing methods can be identified and based on which new method can be developed or existing method can be modified to suit to specific application.
3. To make appropriate interpretation of the results obtained by the software.

1.1. Traditional Optimization Techniques

These are deterministic algorithms with specific rules for moving from one solution to the other. These algorithms have been in use for quite some time and have been successfully applied to many engineering design problems. Examples include non-linear programming, geometric programming, quadratic programming, dynamic programming, generalized reduced gradient method, etc. Although, traditional mathematical programming techniques had been employed to solve optimization problems in manufacturing applications, these techniques have following limitations:

- Traditional techniques do not fare well over a broad spectrum of problem domains.
- Traditional techniques are not suitable for solving multi-modal problems as they tend to obtain a local optimal solution.

- Traditional techniques are not suitable for solving problems involving large number of constraints.

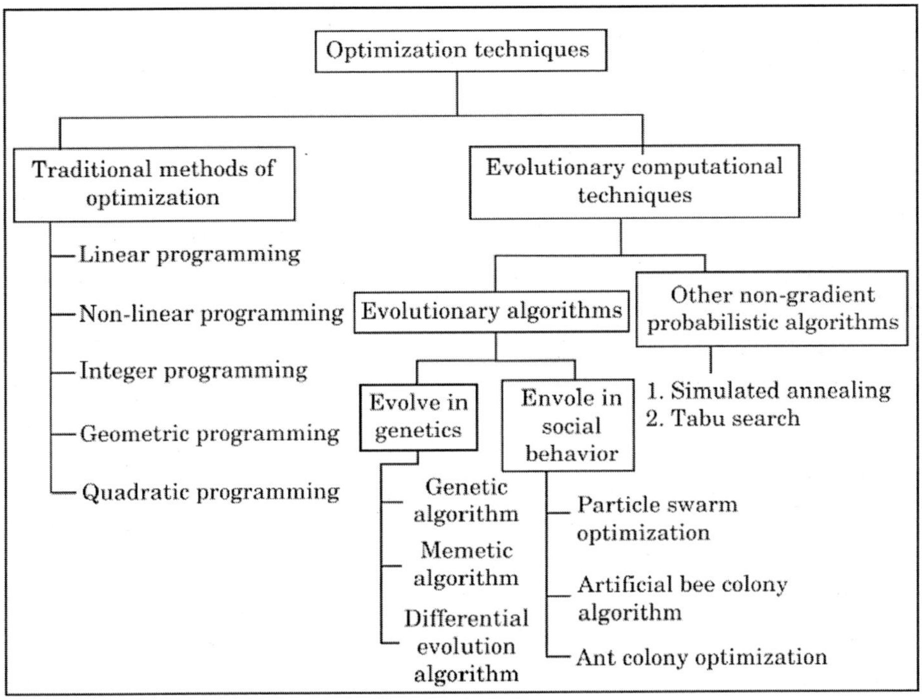

Fig. 1.1: Shows classification of optimization techniques

To understand the concept of multimodality, consider an optimization problem to determine the value of x that minimize $f(x)$ as shown in Fig. 1.2. As the function is multi-modal, there are two minima; local minimum at x = P and global minimum at x = Q. If we apply traditional method of optimization with initial solution at x = A, the algorithm will converge to local optimum solution. This is due to the fact that most of the traditional methods of optimization are gradient based and their search process will be terminated when the gradient becomes zero. However, if we start with initial solution at x = B, then algorithm will converge to global optimum solution. It is quite obvious that, as multi-modality increases, the probability of getting global optimum solution by using traditional optimization methods reduces. The optimum solution obtained by most of the traditional methods is thus highly dependent on the initial solution, which restricts the application of these algorithms to solve unimodal problems only.

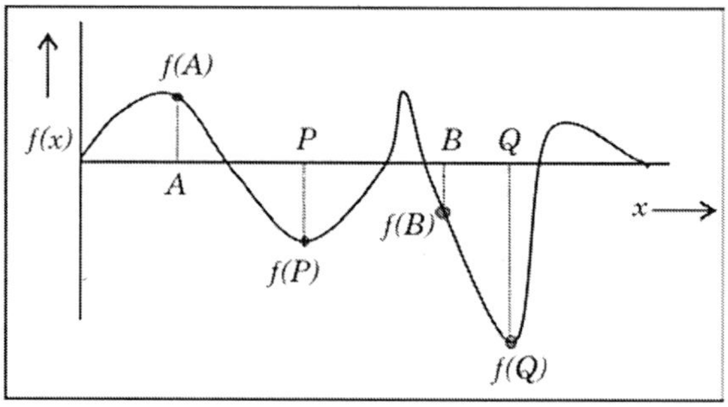

Fig. 1.2: Multimodal function

The ineffectiveness of traditional optimization methods to solve multimodal problems is demonstrated through following case study on optimization of ultrasonic machining process.

1.2. Optimization of Ultrasonic Machining Process

Ultrasonic machining (USM) is a process in which material is removed due to action of abrasive grains. The abrasive particles are driven into the work surface by a tool oscillating normal to the work surface at a high frequency. The tool is shaped as the approximate mirror image of the configuration of cavity desired in work-piece as shown in Fig. 1.3.

The process can handle large variety of materials as it is not limited by the electrical or chemical characteristics of work material. The process has the advantage of machining hard and brittle materials into complex shapes with good accuracy and reasonable surface finish. Considerable economy results from the ultrasonic machining of hard alloy tools and dies on account of their high resistance. The process is particularly suitable to make holes with a curved axis of any shape that can be made on tool. However, the major limitation of the ultrasonic machining process is its low material removal rate characteristic. It is therefore very essential to improve the metal cutting rates for this process without affecting surface finish of the work-piece. This can be achieved through optimum selection of various machining process parameters influencing the material removal rate and surface finish in ultrasonic machining process. The optimization model thus consist of objective function as to maximize the material removal rate subjected to a constraint of surface roughness. The optimization model for ultrasonic machining process proposed by Jain *et al.* (2001) is discussed below:

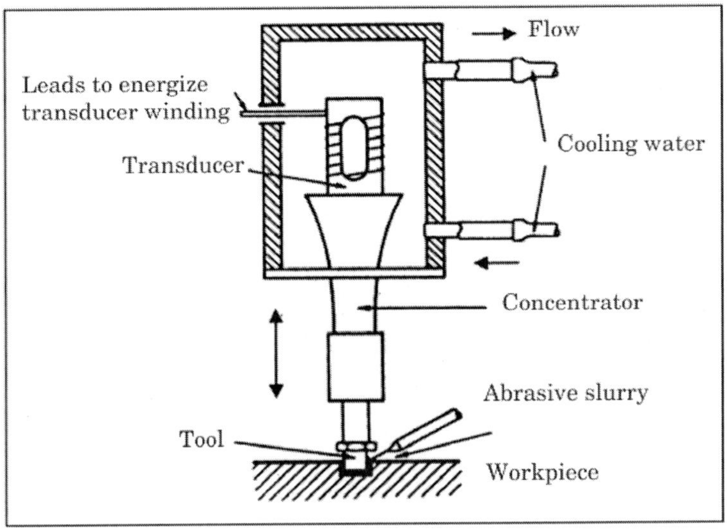

Fig. 1.3: Ultrasonic machining process (Jain and Jain, 2001).

Objective function: Maximization of material removal rate (Z_1) given by Eq. (1.1).

$$Z_1 = \frac{4.963 A_t^{0.25} K_u^{0.75}}{\left[\sigma_{fw}(1+\lambda)^{0.75}\right]} F_s^{0.75} A_v^{0.75} C_{av}^{0.25} d_m f_v \tag{1.1}$$

Where, the decision variables are A_v, amplitude of vibration (mm); f_v frequency of vibration (Hz); d_m Mean diameter of abrasive grain (mm); C_{av}, Volumetric concentration of abrasive particles in slurry, and F_s, static feed force (N). K_u is a constant of proportionality (mm^{-1}) relating mean diameter of abrasive grains, and diameter of projections on an abrasive grain ($= K_u d_m^2$).

Constraint: The constraint is on permissible surface roughness as given below:

$$Z_2 = \frac{1154.7}{\left[A_t \sigma_{fw}(1+\lambda)\right]^{0.5} R_{a_{max}}} \frac{\left[F_s A_v d_m\right]^{0.5}}{C_{av}} \geq 0 \tag{1.2}$$

Variable Bounds

$$0.005 \leq A_v \leq 0.1 \text{ (mm)};$$
$$10000 \leq f_v \leq 40000 \text{ (Hz)};$$

$$0.007 \leq d_m \leq 0.15 \text{ (mm)};$$
$$0.05 \leq C_{av} \leq 0.5;$$
$$4.5 \leq F_s \leq 45 \text{ (N)}.$$

This model is now solved by applying a traditional optimization method using Matlab Optimization toolbox "Optimtool" considering default setting of algorithm specific parameters as shown in Fig. 1.4. Script files (m-files) for objective function and constraint function are 'fun_usm.m' and 'con_usm.m' respectively.

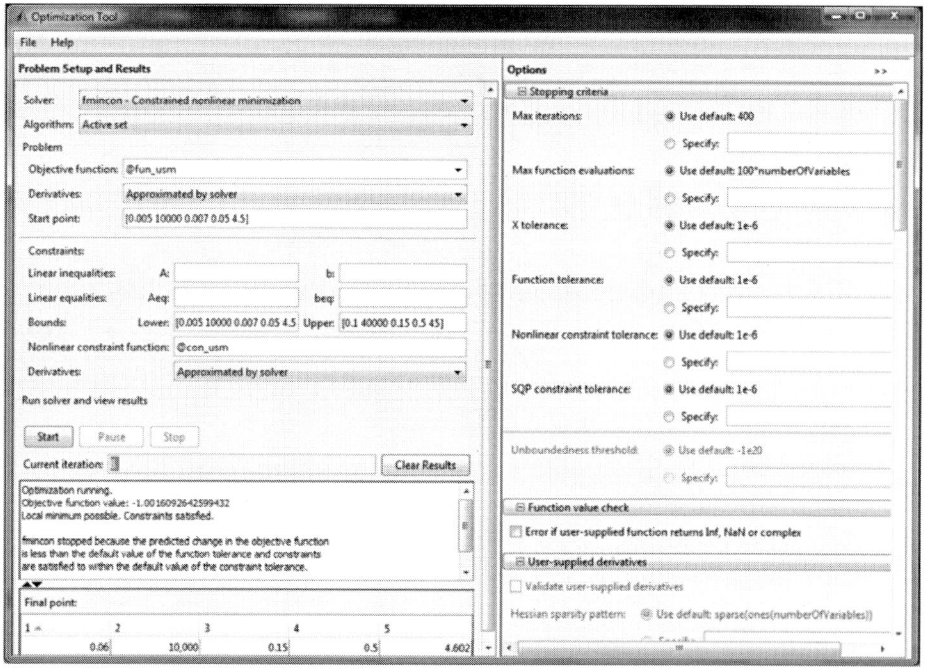

Fig. 1.4: Optimization of USM process using Matlab solver 'fmincon'.

As shown if Fig. 1.4, considering the start point as: $A_v = 0.005$ mm, $f_v = 10000$ Hz, $d_m = 0.007$ mm, $C_{av} = 0.05$, $F_s = 4.5$N, the optimum solution (final point) is $A_v = 0.06$ mm, $f_v = 10000$ Hz, $d_m = 0.15$ mm, $C_{av} = 0.5$, $F_s = 4.602$ N resulting into material removal rate (Z_1) of – 1.001 mm^3/s with surface roughness constraint (Z_2) well satisfied. As we are applying a minimization algorithm ('fmincon' is a solver for constrained non-linear minimization problem) to solve maximization problem, it is necessary to transform the objective function suitably. One of the several ways is to assign a minus sign to the objective function and let the algorithm to minimize it. The optimum solutions after removing minus sign represent the maximum value of objective function. The optimum values of material removal rates (MRR) corresponding to initial solutions considered in five independent runs are reported in Table 1.1. It can be seen that algorithm offers different optimum solutions for different initial solutions. Also, no solution reported in Table 1.1 is a global optimum solution. This clearly indicates the multimodal nature of the optimization model and also how traditional method fails to obtain the global optimum solution.

Table 1.1: Initial solution and corresponding optimum solution obtained by using 'fmincon' command for five independent runs.

Initial solutions					Optimum value of MRR (mm^3/s)
A_v (mm)	f_v (Hz)	d_m (mm)	C_{av}	F_s (N)	
0.01	20000	0.05	0.2	10	2.003
0.007	25000	0.12	0.4	5	2.504
0.1	34000	0.007	0.06	45	3.405
0.1	35000	0.15	0.5	45	3.505
0.008	32000	0.009	0.4	10	3.205

It can be thus concluded that as performance of traditional optimization method is governed by the initial solution, the success rate of these algorithms is relatively low for solving multi modal models.

***Success rate*:** If the algorithm is run for 100 times and able to hit global optimum solution for 5 times, then success rate is 5%.

It is also to be noted that it is not possible to classify the real life model as a unimodal or multimodal and no guidelines are available for selection of the initial solution. Thus, being deterministic in nature, there is a great risk in applying traditional optimization methods to multi-modal optimization models. It is possible to make algorithm performance (in terms of success rate) independent of initial solution, only if the algorithm is probabilistic in nature. As an attempt, to achieve the global optimum solution irrespective of initial solution chosen, another class of optimization methods known as evolutionary computations emerged. Evolutionary computations is a family of algorithms for global optimization based on biological, molecular or neurological phenomenon that mimics the metaphor of natural biological evolution and/or the social behavior of species. In technical terms, they are a family of population-based problem solvers with a metaheuristics or stochastic optimization character. These algorithms are:

a) Evolutionary algorithms which evolve genetically such as genetic algorithms (GA), memetic algorithm (MA), and differential evolution algorithm (DEA) etc.
b) Evolutionary algorithms which evolve in their social behavior. Most of the swarm intelligence based algorithms such as artificial bee colony (ABC) algorithm, shuffled frog leaping (SFL) algorithm, ant colony optimization (ACO) algorithm, particle swarm optimization (PSO), civilized swarm optimization (CSO), firefly algorithm, etc. falls into this category.
c) Non-gradient probabilistic algorithms such as simulated annealing algorithm (SA), tabu search (TS), cuckoo search algorithm etc.

Evolutionary computational techniques are usually population based in which set of initial solutions (population) is generated and updated iteratively. Each new generation is produced by stochastically removing less desired solutions, and introducing small random changes. As a result, the population will gradually evolve to increase in fitness. Evolutionary computational techniques can produce highly optimized solutions in a wide range of problem settings. Several variants have been proposed by researcher to suit to particular class of application. Evolutionary computational techniques are also termed as soft computing technique. Soft computing techniques are essentially having few parameters (referred as algorithm specific parameters) which can be adjusted or tuned to suit to specific application. As an example in case of genetic algorithm, algorithm specific parameters are crossover probability, mutation probability, string length etc., for particle swarm optimization, inertia weight, cognitive leaning parameter, social learning parameter etc. These algorithm specific parameters must be chosen so as to properly balance between exploration and exploitation to avoid premature convergence to a local optimal solution yet still ensure a good rate of convergence to the optimum. It is therefore essential to understand the convergence behavior of these algorithms to tune the algorithm specific parameters and then only the real beauty of these algorithms can be realized.

1.3. Applications of Evolutionary Computational Methods to Manufacturing Systems

In manufacturing systems there are several issues which can be well handled by evolutionary computational methods. Few of them are listed below.

1.3.1. *Dynamic scheduling problem*

The choice of dynamic or static scheduling is an important one and has far-reaching consequences for the system. Static scheduling is a good fit with a time-triggered design, and dynamic scheduling is a good fit for an event-triggered design. Static scheduling must be carefully planned in advance, with considerable effort going into choosing the various parameters. Dynamic scheduling does not require as much advance work, since scheduling decisions are made on-the-fly, during execution.

Dynamic scheduling can potentially make better use of resources than static scheduling. In the latter, the system must frequently be over dimensioned to have so much capacity that it can handle even the most unlikely cases. However, in a hard real-time system, wasting resources is often the price that must be paid to guarantee that all deadlines will be met. In dynamic scheduling problems, a stream of jobs is arriving stochastically over time. Each job requires a combination of resources, sequentially and/or in parallel, for different processing times. The existence of any particular job and its corresponding characteristics is not known until its arrival. Solving dynamic scheduling problems is challenging both due to the combinatorics of the interaction of jobs, resources, and time, and due to the stochastics: to make a decision, one can use only the information that is known with certainty at a decision point and the stochastic properties of scenarios that may occur in the future (Dimopoulos and Zalzala, 2000).

Therefore, until recently, evolutionary computation methods have rarely been used for dynamic scheduling, due to their inability to cope with real-time decision making. They were developed and tested on static scheduling problems that did not require real-time control. However, in the last few years, evolutionary algorithms have been employed as parts of hybrid dynamic scheduling systems, which exploit their useful characteristics.

1.3.2. *Material flow optimization*

In a loading problem of flexible manufacturing system (FMS), part type selection and operations allocation are two critical problems. The total completion time of a product for the selected process plan in the system can be minimum for the loading problem. But, in a real time scheduling system, this process plan may not be optimum because of consideration of waiting time of product and machine. So, the total completion time and thereby the material flow of the selected process plan in the FMS may be high. To overcome this issue, an integrated approach of part type selection and an operation allocation problem *i.e.*, production planning problem and scheduling problem can be considered to optimize material flow in FMS. Loading and scheduling problems are NP-hard in nature. To deal with this complexity, evolutionary optimization methods can be applied.

1.3.3. *Process planning*

With the advent of computer technologies, there is a general demand for computer-aided process planning (CAPP) systems to assist human planners and achieve the integration of computer-aided design (CAD) and computer aided manufacturing (CAM). Moreover, the introduction of new manufacturing technologies, such as computer-integrated manufacturing systems (CIMS) and design for manufacturing (DFM), causes an even greater demand for automated CAPP systems. Computer aided process planning involves: (i) automatic selection of operations including selection of machine tool, cutting tools, tools etc. based on the feature geometry and available machining resources and (ii) automatic sequencing of operation maintaining the precedence relationship among all operations (Zhang *et al.,* 1997). Evolutionary algorithm can provide a valid option for solving above mentioned issues in process planning.

1.3.4. *Job shop and Flow shop scheduling problem*

The job-shop scheduling problem (JSSP) consists of ordering jobs to be processed in machines. Each job involves a number of different machining operations. The following conditions hold for the classic formulation of the JSSP (Dimopoulos and Zalzala, 2000): (a) each machine can process only one job at a time (b) the sequence of operations for each job is predefined (c) two operations of the same job cannot be processed at the same time (d) preemption is not allowed (an operation cannot be withdrawn from a machine unless it is completed) (e) processing times are known in advance (f) transportation time between machines is zero. The quality criterion most often used for the JSSP is minimization of make span which is defined as the completion time of the final job to leave the system.

1.3.5. *Cellular manufacturing and machine layout*

Cellular Manufacturing is an application of Group Technology, in which similar machines or processes have been grouped into cells, which allows taking advantage of the similarity between parts or machines, through standardization and common processing like assembly, machining, tooling, jigs – fixtures or mounting etc. Cellular manufacturing system leads to shorten the manufacturing lead time, restricts inter cellular movements up to large extent and reduce setup time. Cellular Manufacturing is most suitable for batch manufacturing. The cellular manufacturing is intended to have the flexibility to produce a high variety of low demand products, while maintaining the high productivity on large scale production. Cellular manufacturing attempts to simplify the multi-objective / multifaceted problem. It leads to accomplishment of following objectives (Arora *et al.*, 2013): (a) To shorten manufacturing lead time. (b) Minimization of intercellular movement of material. (c) Minimization of throughput times. (d) To reduce setup times. (e) To reduce work- in - process inventories. (f) Maximization of labour and machine utilization. (g) Maximization of output and to improve quality. (h) Minimization of operating cost.

1.3.6. *Parametric optimization of manufacturing processes*

In today's manufacturing environment, many large industries have attempted to introduce the highly automated and computer-controlled machines as their strategy to adapt to the ever-changing competitive market requirement. Due to high capital and machining costs, there is an economic need to operate these machines as efficiently as possible in order to obtain the required pay back. The success of the machining operation depends on the selection of machining process parameters. Determination of optimum process parameters of any machining process is usually a difficult work where the following aspects are required: knowledge of manufacturing process, empirical equations to develop realistic constraints, specification of machine tool capabilities, development of effective optimization criteria, and knowledge of mathematical and numerical optimization techniques. A human process planner selects proper machining process parameters using his own experience or from the handbooks. But these parameters do not give optimal result. The selection of optimum process parameters using evolutionary computational methods play a significant role to ensure quality of product, to reduce the machining cost, to increase productivity in computer-controlled machining processes and to assist in computer-aided process planning.

1.3.7. *Assembly sequence planning problem*

It is often the case that several products with similar characteristics (models) are assembled in a single line (mixed-model assembly lines). The sequencing of models in mixed-model assembly lines is an important task, especially if we wish to apply the JIT principle in the production line. There are a number of objectives associated with this task, like the minimization of the line's length, the minimization of total utility work, and the minimization of the variability of parts' consumption.

1.3.8. *Assembly line balancing*

Another well-known optimization problem of assembly lines is the assembly line balancing problem. Given workstations and parts to be assembled, the assignment of parts to workstations should be defined according to certain optimization criteria. Two versions of the problem are usually considered. The first version aims to minimize the total number of workstations in the plant given a fixed cycle time, while the second version aims to minimize the cycle time, given a fixed number of workstations. Secondary objectives like the minimization of balance delay and the minimization of probability of line stoppage can also be considered.

1.3.9. *Tool and die design*

There are several optimization issues related to tool and die design in manufacturing industries such as: Blank nesting, spring back elimination, clearance optimization etc. in case of press tool design, deterministic location and clamping system in case of fixture design, parting line selection in forging die design, determination of optimum gate location, cooling system design in die casting dies and injection molds.

1.3.10. *Tool path planning in hole making operations*

Optimization of hole-making operations in manufacturing industry plays a vital role. Tool travel and tool switch planning are the two major issues in hole-making operations. Many industrial applications such as moulds, dies, engine block, automotive parts etc. requires machining of large number of holes. Large number of machining operations like drilling, enlargement, tapping or reaming are required to achieve the final size of individual hole, which gives rise to number of possible sequences to complete hole-making operations on the part depending upon the location of hole and tool sequence to be followed. It is necessary to find the optimal sequence of operations which minimizes the total processing cost of hole-making operations.

Application of evolutionary computational methods to various such applications related to manufacturing systems are discussed in detail in subsequent chapters.

2

Genetic Algorithm

2.1. Introduction

The last few years have seen important advances in the use of genetic algorithms to address challenging optimization problems related to manufacturing systems. The main idea of genetic algorithm (GA) is to mimic the natural selection and the survival of the fittest. In GA, the solutions are represented as chromosomes. The chromosomes are evaluated for fitness values and they are ranked from best to worst based on fitness value. The process to produce new solutions in GA is mimicking the natural selection of living organisms, and this process is accomplished through repeated applications of three genetic operators: selection, crossover, and mutation (Holland, 1975; Goldberg, 1989). Flowchart of a basic genetic algorithm is shown in Fig. 2.1 below.

Genetic Algorithm differs from traditional optimization techniques with respect to the following:

1. GA work with a coding of the parameter set, not the parameters themselves.
2. GA search from a population of points and not a single point.
3. GA use information of fitness function and not derivatives or other auxiliary knowledge
4. GA use probabilistic transitions rules and not deterministic rules.
5. It is very likely that the expected GA solution will be global solution.

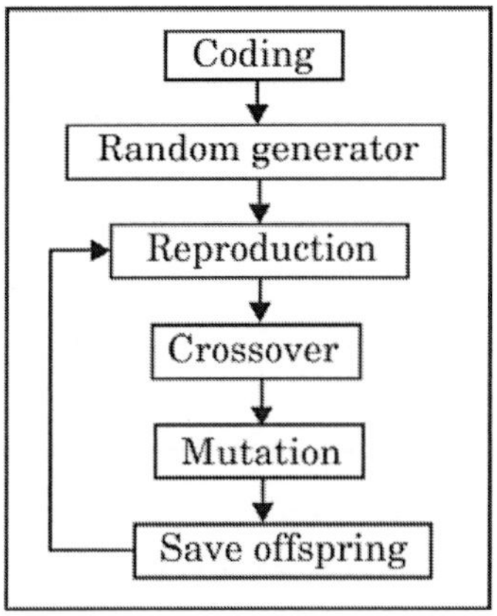

Fig. 2.1: Flowchart of genetic algorithm (Goldberg, 1989).

2.2. Mechanism of Working of Genetic Algorithm (GA)

The mechanism of the three basic operators in the genetic algorithm: reproduction, crossover and mutation are demonstrated through a simple example discussed below.

Consider an objective function having two variables x_1 and x_2:

$$Minimize\ z = x_1^2 - 3x_2 + 10 \text{ such that } 0 \leq x_{1,}x_2 \leq 3 \qquad (2.1)$$

It can be easily seen that to minimize this function, x_1 should be minimum and x_2 should be maximum. Thus, the optimum solution is $x_1 = 0$ and $x_2 = 3$ with optimum function value $(z) = 1$. Now, we shall see how the steps of basic genetic algorithm can be implemented to solve this problem.

Step 1: *Generate randomly initial population of solutions*

As genetic algorithm is population based algorithm, it starts with number of initial solutions (population) generated randomly unlike a single initial solution in case of traditional methods of optimization. The decision on number of initial solutions is problem specific. Let us consider five initial solutions as shown in Table 2.1 for solving this problem.

Table 2.1: Random generation of initial solutions.

Sl. no.	*Substring (x_1)* Coded value	*Substring (x_1)* Decoded value	*Substring (x_2)* Coded value	*Substring (x_2)* Decoded value	x_1	x_2	z
1	1001	9	0010	2	1.8	0.4	12.04
2	0101	5	0001	1	1	0.2	10.4
3	0010	2	1110	14	0.4	2.8	1.76
4	0100	4	1100	12	0.8	2.4	3.44
5	1100	12	1001	9	2.4	1.8	10.36

In binary coded version of GA, solutions are generated in binary coded form and not in real form. The total string length is divided into number of substrings one for each variable. As shown in Table 2.1, the sub-string length for both variables, x_1 and x_2 is considered as 4. However, it is not necessary that the sub-string length for all variables should be the same. It depends on expected accuracy of a particular variable which is determined by using a formula: $\frac{x_{max} - x_{min}}{2^n - 1}$ .Where, n is a sub-string length for variable x. Thus for n = 4, accuracy of the variable is $\frac{3}{15} = 0.2\,units$. Higher accuracy can be achieved by considering higher value of n. However, with increase in the value of n, the computational efforts required to solve the problem are also increased. If we consider substring length of 4 for both variables (x_1 and x_2), the string length will become 8. The decoded value for each sub-string is obtained by applying the rules of binary decoding. Thus for coded value of 1001 the decoded value is $1\times2^3 + 0\times2^2 + 0\times2^1 + 1\times2^0 = 9$. However, the decoded value does not represent the actual value of variable. The actual value of variable is obtained by using formula: $x = x_{min} + \frac{x_{max} - x_{min}}{2^n - 1} \times decoded\ value$ (Deb, 2004). Thus for sub-string 1001, the actual value of variable is: $x = 0 + 0.2\ 0215 \times 9 = 1.8$. It is to be noted that even though binary coded solutions are randomly generated, in no case the corresponding values of variables will exceed their bounds. For minimum possible binary value of 0000, corresponding value of variable is 0 and for maximum possible binary value of 1111, corresponding value of variable is 3. Thus for any binary number generated between 0000 and 1111, the corresponding values of variables will lie between their bounds 0 and 3. The function values (z) are then obtained by substituting actual values of variables in Eq. (2.1).

Step 2: *Reproduction*

As GA works on the principle of survival of the fittest, the strong solutions in the population will survive and weak solutions will be eliminated from the process. This process of natural selection is carried out in the reproduction stage. Thus in reproduction stage, fit solutions are copied for more number of times (proportional to their fitness values) and weak solutions will be eliminated as shown in Table 2.2.

Table 2.2: Reproduction

Sl. no.	x_1	x_2	*z*	*Fitness value strength*	*Individual strength*	*Relative strength*	*Cumulative*	*Rand. no*	*Copied string* x_1	*Copied string* x_2	*Z*
1	1.8	0.4	12.04	0.083	0.365	0.073	0.0731	0.440	0010	1110	1.76
2	1	0.2	10.4	0.096	0.423	0.084	0.157	0.220	0010	1110	1.76
3	0.4	2.8	1.76	0.568	2.503	0.500	0.658	0.707	0100	1100	3.44
4	0.8	2.4	3.44	0.290	1.280	0.256	0.914	0.614	0010	1110	1.76
5	2.4	1.8	10.36	0.096	0.425	0.085	1.000	0.669	0100	1100	3.44

The principle of survival of the fittest makes the genetic algorithm suitable solving maximization problems. Thus for GA, solution 1 shown in Table 2.1 is best as it has maximum function value. However, as our problem is of minimization in nature, solution 3 is best. Hence to proceed further with the steps of GA, it is necessary to apply appropriate transformation and convert the objective function value into corresponding fitness value of the solution. One way of doing this is to take the reciprocal of the objective function values. The reciprocal of objective function is referred as fitness function. So by considering this fitness function value, solution 3 is best for GA also. As shown in Table 2.2, the strength of individual solution is obtained by dividing its fitness function value by average fitness function value of the population and the relative strength is obtained by dividing the individual strength by population size. The cumulative relative strength of last solution in the population (solution 5 in this case) will be always one. Although there are several selection operators in GA, roulette wheel selection is presented here.

The cumulative relative strengths are plotted in the form of roulette wheel as shown in Fig. 2.2. Now, if a random number is generated between 0 and 1, there is maximum chance that this random number will lie in the region between 0.157 and 0.658 followed by the region between 0.658 and 0.914. As shown in Table 2.2, let the first number generated randomly is 0.440, will lie in the region 0.157 to 0.658. Since this region represents the contribution of solution 3, the solution 1 is replaced by solution 3. Thus, at the end of reproduction, solutions 1, 2 and 4 are replaced by solution 3 and solutions 3 and 5 are replaced by solution 4. Thus solution 3 is copied for 3 times and solution 4 is copied for 2 times. Being inferior, solutions 1, 2 and 5 are eliminated in this process. It is to be noted that population at the end of reproduction consists of solutions from initial population only but with strong solutions copied for multiple times depending on their fitness values and weak solutions are eliminated.

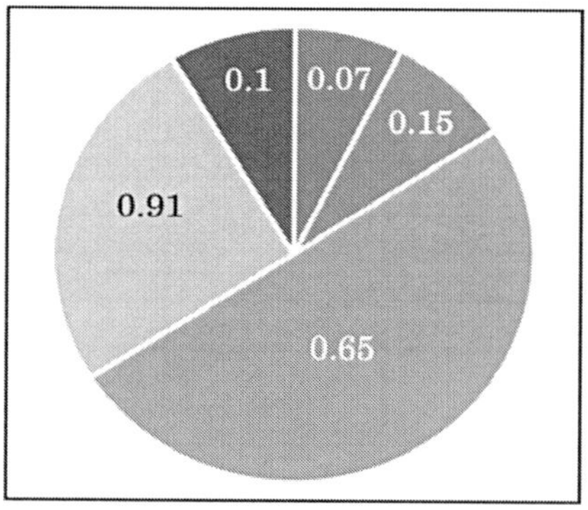

Step 3: *Crossover*

Crossover refers to the blending of chromosomes from the parents to produce new chromosomes for the offspring. The genetic algorithm selects two strings at a random from the mating pool. It is then decided whether to crossover using a parameter called crossover probability. If the crossover takes place then a random slicing point is chosen in the string. The sliced regions are then mixed to create two new strings. The crossover and mutation processes are presented in Table 2.3.

Single point crossover is presented in Table 2.3. Let, the candidate solution for crossover is solution 3 and solution 4 and randomly selected crossover site is 2. In crossover process, since the crossover site is 2, in the given string of 8 bits, first two bits (genes) of both solutions 3 and solution 4 will remain same and remaining bits (bits 3 to 8) will be exchanged with those of another solution. Thus two new solutions (child strings) are formed as 01101110 and 00001100. In fact when population size is large (say 50), one has to choose appropriate crossover probability, usually between 0.7–0.9. If crossover probability is considered as 0.8, then 80% of the total solutions (= 40 solutions) will undergo crossover. It means 20 pairs of solutions are chosen randomly for crossover.

Table 2.3: Crossover and mutation

Sl. no.	***Reproduction***			***Crossover***		***Mutation***		***Solution(after first generation)***		
	x_1	x_2	***Site***	x_1	x_2	x_1	x_2	x_1	x_2	z
1	0010	1110		0010	1110	0000	0110	0	1.2	6.4
2	0010	1110		0010	1110	0010	1111	0.4	3.0	1.16
3	0100	1100	2	0110	1110	0110	0110	1.2	1.2	7.84
4	0010	1110	2	0000	1100	0001	1100	0.2	2.4	2.84
5	0010	1110		0010	1110	0010	1111	0.4	3.0	1.16

Step 4: *Mutation*

Although crossover can generate a staggering amount of different strings, there may not be enough variety of stings to ensure that the entire problem is covered. This may lead to converging on strings that are not quite close to the optimum it seeks. To overcome this problem a mutation operator is introduced into a genetic algorithm. For each string element in each string in the mating pool the algorithm checks to see if it should perform mutation and if it should, the string element is flipped (in case of binary strings). The mutation thus helps to prevent the population from stagnating and maintains the diversity throughout the iterations. However, the mutation probability should be kept very low as a high mutation rate will destroy fit strings and degenerate the algorithm into a random walk. As shown in Table 2.3, the bits marked bold are candidates for crossover and these bits are flipped (mutated) *i.e.,* 1 becomes 0 and 0 becomes 1. The solutions at the end of mutation are presented in Table 2.3. These solutions in coded form are converted to actual solutions as discussed in step 1. It is observed that in the updated population, the best value is 1.16 which is better than previous best (=1.76). This completes first iteration. Now the steps 1 to 4 are repeated till solution with desired accuracy is obtained and the algorithm run is terminated.

Step 5: *Termination*

The termination condition of a Genetic Algorithm is important in determining when a GA run will end. It has been observed that initially, the GA progresses very fast with better solutions coming in every few iterations, but this tends to saturate in the later stages where the improvements are very small. We usually want a termination condition such that our solution is close to the optimal, at the end of the run.

The various termination criteria are (www.mathworks.com):

- ***Generation*:** The algorithm stops when the number of generations reaches the value of Generations.
- ***Time limit*:** The algorithm stops after running for an amount of time in seconds equal to Time limit.
- ***Fitness limit*:** The algorithm stops when the value of the fitness function for the best point in the current population is less than or equal to Fitness limit.
- ***Stall generations*:** The algorithm stops when the average relative change in the fitness function value over Stall generations is less than Function tolerance.
- ***Stall time limit*:** The algorithm stops if there is no improvement in the objective function during an interval of time in seconds equal to Stall time limit.
- ***Stall test*:** The stall condition is either average change or geometric weighted. For geometric weighted, the weighting function is 1/2n, where n is the number of generations prior to the current. Both stall conditions apply to the relative change in the fitness function over Stall generations.
- ***Function tolerance*:** The algorithm runs until the average relative change in the fitness function value over Stall generations is less than Function tolerance.
- ***Constraint tolerance*:** The Constraint tolerance is not used as stopping criterion. It is used to determine the feasibility with respect to nonlinear constraints.

After this detail discussion about working of GA, various real life application case studies related to manufacturing systems are discussed in following sections.

2.3. Optimization of the Plant Layout in Production of an Automobile Transmission System

An important aspect of organizational efficiency is the plant layout, defined as the mechanism which involves knowledge of the space requirements for the facilities and also involves their proper arrangement so that continuous and steady movement of the production cycle takes place. Proper arrangement of facilities reduces material handling distance, cost and equipment used in the production process. A case study of a manufacturing plant producing components of an automobile transmission system is discussed in this section to demonstrate the effectiveness of genetic algorithm to solve plant layout optimization problem. An automobile transmission system is a part of machine that consists of a power source and a power transmission system, which provides controlled application of the power. The optimization of layout of production means can be carried out in following four steps:

Step 1. ***Description of manufacturing process:***

The process consists of manufacturing of following three parts:

(a) ***Yoke shaft (propeller shaft)*****:** Raw material for propeller shaft is a forged mild steel bar having hardness 230 to 260 BHN. The quantity to be produced is 800 parts per week. The sequence of operations to be followed for processing of propeller shaft is as given below:

Sl. no.	*Process*	*Machine code*	*Location*
1	Cutting	M1	L1
2	Centering/Facing	M2	L2
3	Cross hole drilling	M3	L3
4	Finish turning	M4	L4

(b) ***Mid ship shaft*****:** Raw material for mid ship shaft is a forged steel bar having hardness 230 to 260 BHN. The quantity to be produced is 1500 parts per week. The sequence of operation to be followed for processing of mid ship shaft is as given below.

Sl. no.	*Process*	*Machine code*	*Location*
1	Centring	M5	L5
2	Centring	M5	L5
3	Rough turning	M4	L4

Sl. no.	*Process*	*Machine code*	*Location*
4	Rough turning	M6	L6
5	Centre reaming	M5	L5
6	Finish turning	M6	L6

(c) ***Sleeves*:** Raw material for sleeve is a rolled bar having hardness 230 to 260 BHN. The quantity to be produced is 3500 parts per week. The sequence of operation to be followed for processing of sleeve is as given below.

Sl. no.	*Process*	*Machine code*	*Location*
1	Cutting	M1	L1
2	Centring	M7	L7
3	Turning	M8	L8
4	First side drilling	M10	L10
5	Second side drilling	M9	L9
6	Hardening	M14	L14
7	Finish Turning	M6	L6
8	Second side boring	M11	L11
9	First side boring	M12	L12
10	Finish total length	M6	L6
11	Drilling	M13	L13

Step 2: *Description of the existing layout in the company*

The plant thus has total 14 machines/facilities M1 to M13, and Heat treatment facility (M14). All these machines can be relocated except

Table 2.4: Distance matrix for machine locations.

	RM	L_1	L_2	L_3	L_4	L_5	L_6	L_7
RM	0	3.9	6.3	8.7	11.1	11.1	8.7	6.3
L_1	3.9	0	0.5	2.91	5.32	5.92	5.42	3.02
L_2	6.3	0.5	0	0.5	2.41	3.01	2.51	0.6
L_3	8.7	2.91	0.5	0	0.5	1.1	0.6	1.1
L_4	11.1	5.32	2.41	0.5	0	0.6	1.1	3.51
L_5	11.1	5.92	3.01	1.1	0.6	0	0.5	2.91
L_6	8.7	5.42	2.51	0.6	1.1	0.5	0	0.5
L_7	6.3	3.02	0.6	1.1	3.51	2.91	0.5	0
L_8	10.3	12.43	14.84	17.25	19.66	19.66	17.25	14.84
L_9	13.33	14.84	17.24	19.66	22.07	22.07	19.66	17.24
L_{10}	10.73	12.23	14.64	17.05	19.46	19.46	17.06	14.64
L_{11}	14.13	15.63	18.14	20.55	22.96	22.96	20.55	18.14
L_{12}	16.53	18.03	20.55	22.96	25.37	25.37	22.96	20.55
L_{13}	18.94	20.43	22.96	25.37	27.78	27.78	25.37	22.96
L_{14}	0	3.9	6.3	8.7	11.1	11.1	8.7	6.3

Table 2.4: (*Contd...*)

	L_8	L_9	L_{10}	L_{11}	L_{12}	L_{13}	L_{14}
RM	10.3	13.33	10.73	14.13	16.53	18.94	0
L_1	12.43	14.83	12.23	15.63	18.03	20.43	3.9
L_2	14.84	17.24	14.64	18.14	20.55	22.96	6.3
L_3	17.25	19.66	17.05	20.55	22.96	25.37	8.7
L_4	19.66	22.07	19.46	22.96	25.37	27.78	11.1
L_5	19.66	22.07	19.46	22.96	25.37	22.78	11.1
L_6	17.25	19.66	17.06	20.55	22.96	25.37	8.7
L_7	14.84	17.24	14.64	18.14	20.55	22.96	6.3
L_8	0	0.5	0.4	3.8	6.21	8.62	10.3
L_9	0.5	0	2.81	6.21	8.62	11.03	13.33
L_{10}	0.4	3.01	0	0.5	291	5.32	10.73
L_{11}	3.8	6.21	0.5	0	0.5	2.41	14.13
L_{12}	6.21	8.62	2.91	0.5	0	0.5	16.53
L_{13}	8.62	11.03	532	2.41	0.5	0	18.94
L_{14}	10.3	13.33	10.73	14.13	16.53	18.94	0

that for heat treatment. The existing plant layout with these 14 machines is as shown in Fig. 2.3. The distance of one location to other location is shown in Table 2.4.

Step 3: *Optimization of plant layout using genetic algorithm*

The objective function for this problem (Z) is to determine appropriate location for each machine so as to minimize the total distance travelled by the in-process material. The objective function is formulated as:

$Z = \Sigma Z_i$, where $i = 1, 2, 3$ (2.2)

Z_1 = Distance travelled: Raw material–M1–M2–M3–M4–Dispatch

Z_2 = Distance travelled: Raw material–M5–M4–M6–M5–M6–Dispatch

Z_3 = Distance travelled: Raw material–M1–M7–M8–M10–M9–M14–M6–M11–M12–M6 M13–Dispatch

***Constraint*:** The constraint is on location of heat treatment machine (M14) which is to be fixed at location 14.

Real coded genetic algorithm (RCGA) is implemented for optimization of plant layout in the presented case study. The use of real parameters in RCGA makes it possible to use large domains (even unknown domains) for the variables, which is difficult to achieve in binary implementations where increasing the domain would mean sacrificing precision, assuming a fixed length for the chromosomes. Another

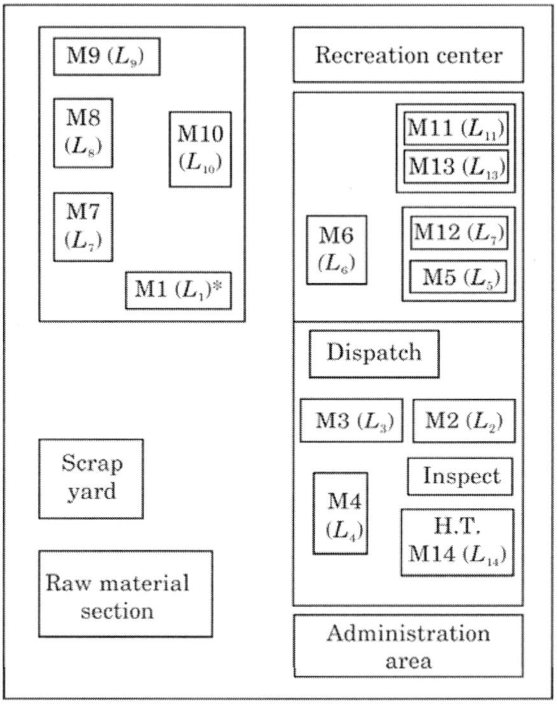

Fig. 2.3: Plant layout (before optimization) for production of an automobile automobile transmission system. (*'*Li*' indicates i^{th} location)

advantage when using real parameters is their capacity to exploit the graduality of the functions with continuous variables, where the concept of graduality refers to the fact that slight changes in the variables correspond to slight changes in the function. RCGA has the capacity for the local tuning of the solutions. There are genetic operators such as uniform mutation that allows the tuning to be produced in a more suitable and faster way than in the binary coded GA's, where the tuning is difficult because of the Hamming cliff effect. Using real coding, the representation of the solutions is very close to the natural formulation of many problems, *e.g.,* there are no differences between the genotype (coding) and the phenotype (search space). Therefore, the coding and decoding processes that are needed in the binary coded GA's are avoided, thus increasing the GA's speed. The steps of implementation of RCGA to the problem formulated are given below:

i) Generation of Initial random solutions

Considering the population size as 10, machines are randomly allocated to each location L_i to get 10 different layouts as shown in Table 2.5. The objective function is evaluated using Eq. (2.2).

Table 2.5: Initial randomly generated layouts.

Sl. no.	L_1	L_2	L_3	L_4	L_5	L_6	L_7	L_8	L_9	L_{10}	L_{11}	L_{12}	L_{13}	L_{14}	Z
1	13	8	2	12	3	7	6	5	11	9	4	14	10	1	407.37
2	7	13	1	6	8	14	3	10	12	2	9	5	11	4	462.72
3	14	10	6	11	9	2	13	8	1	4	12	3	7	5	399.38
4	3	5	11	7	1	4	6	10	14	12	13	8	9	2	365.93
5	8	10	5	13	9	2	6	1	4	12	11	3	14	7	411.07
6	4	7	11	5	1	3	6	8	14	12	9	13	2	10	414.10
7	4	13	3	8	10	5	9	1	6	12	14	7	11	2	400.42
8	1	6	2	8	11	7	12	4	10	14	9	3	13	5	357.04
9	2	13	14	5	1	8	4	6	9	12	7	11	3	10	459.89
10	6	14	10	7	3	1	9	5	2	11	8	13	12	4	417.61

ii) *Selection*

Roulette wheel method is used for natural selection as shown in Table 2.6.

Table 2.6: Selection

Z	***Fitness strength***	***Relative strength***	***Cumulative relative strength***	***Random number***	***Selected solution***
407.37	0.00245	0.09992	0.09992	0.038224	1
462.72	0.00216	0.087968	0.187888	0.845195	9
399.38	0.0025	0.101919	0.289807	0.591626	6
365.93	0.00273	0.111236	0.401043	0.793912	8
411.07	0.00243	0.099021	0.500064	0.215168	3
414.10	0.00241	0.098296	0.598361	0.399171	4
400.42	0.0025	0.101655	0.700015	0.235874	3
357.04	0.0028	0.114005	0.814021	0.847832	9
459.89	0.00217	0.088509	0.90253	0.687406	7
417.61	0.00239	0.09747	1	0.189239	3

Allocation of machines to locations for selected solutions are provided in Table 2.7.

Table 2.7: Allocation of machines to locations for selected solutions.

Selected solution	L_1	L_2	L_3	L_4	L_5	L_6	L_7	L_8	L_9	L_{10}	L_{11}	L_{12}	L_{13}	L_{14}
1	13	8	2	12	3	7	6	5	11	9	4	14	10	1
9	2	13	14	5	1	8	4	6	9	12	7	11	3	10
6	4	7	11	5	1	3	6	8	14	12	9	13	2	10
8	1	6	2	8	11	7	12	4	10	14	9	3	13	5
3	14	10	6	11	9	2	13	8	1	4	12	3	7	5
4	3	5	11	7	1	4	6	10	14	12	13	8	9	2
3	14	10	6	11	9	2	13	8	1	4	12	3	7	5
9	2	13	14	5	1	8	4	6	9	12	7	11	3	10
7	4	13	3	8	10	5	9	1	6	12	14	7	11	2
3	14	10	6	11	9	2	13	8	1	4	12	3	7	5

iii) *Crossover and mutation*

Considering the crossover probability of 80%, 8 solutions (*i.e.*, 4 pairs) will undergo crossover process. These solutions are selected randomly and results of implementation of single point crossover at randomly selected crossover sites are presented in Table 2.8. Table 2.9 shows the objective function value (Z) after mutation.

Table 2.8: Crossover

Cross-over site	*Candi-date sol.*	L_1	L_2	L_3	L_4	L_5	L_6	L_7	L_8	L_9	L_{10}	L_{11}	L_{12}	L_{13}	L_{14}
4	1	1	8	2	6	11	7	13	4	10	14	9	3	12	5
	8	4	6	2	11	1	3	5	8	14	12	9	13	7	10
7	9	2	13	6	5	1	7	4	10	11	12	14	8	9	3
	4	3	5	13	7	1	4	6	8	9	12	14	11	2	10
9	6	2	7	11	5	1	3	6	8	14	4	12	9	10	13
	3	14	3	6	11	4	2	5	8	1	12	9	13	7	10
1	3	4	13	3	8	10	5	9	1	6	12	14	7	11	2
	7	4	10	6	11	9	2	13	8	1	14	12	3	7	5
	3	14	10	6	11	9	2	13	8	1	4	12	3	7	5
	9	2	13	14	5	1	8	4	6	9	12	7	11	3	10

It can be seen from Table 2.9 that at the end of the first iteration best function value is 329.5 which is better than that of initial population (=357.04). The average fitness function value is also improved.

(iv) *Repeat steps (i) to (iii) till convergence*

The steps (i) to (iii) are repeated for 50 iterations. The solution obtained at the end of 50th iteration is shown in Table 2.10.

Table 2.9: Mutation

Mutation site		*Sol.*	L_1	L_2	L_3	L_4	L_5	L_6	L_7	L_8	L_9	L_{10}	L_{11}	L_{12}	L_{13}	L_{14}	*Z*
2	8	1	1	4	2	6	11	7	13	8	10	14	9	3	5	12	429.7
4	12	2	4	6	2	13	1	3	5	8	14	12	9	11	10	7	329.5
2	11	3	2	14	6	5	1	7	4	10	11	12	13	8	3	9	387.8
5	10	4	3	5	13	7	12	4	6	8	9	1	14	11	10	2	378.2
2	10	5	2	4	11	5	1	3	6	8	14	7	12	9	13	10	385.2
2	12	6	14	13	6	11	4	2	5	8	1	12	9	3	10	7	388.4
4	11	7	4	13	3	14	10	5	9	1	6	12	8	7	2	11	421.8
3	10	8	4	10	14	11	9	2	13	8	1	6	12	3	5	7	457.3
1	9	9	1	10	6	11	9	2	13	8	14	4	12	3	5	7	450.0
5	11	10	2	13	14	5	7	8	4	6	9	12	1	11	10	3	433.7

Table 2.10: Result of optimization for plant layout problem.

Location	*Machine code*	*Location*	*Machine code*
L_1	M3	L_8	M1
L_2	M4	L_9	M2
L_3	M6	L_{10}	M7
L_4	M12	L_{11}	M9
L_5	M11	L_{12}	M10
L_6	M13	L_{13}	M8
L_7	M5	L_{14}	M14

The proposed optimum layout is shown in Fig. 2.4.

Comparison of total movement of yoke shaft material for existing layout and optimum layout is presented in Table 2.11.

Table 2.11: Total movement of yoke shaft material for existing and optinum layout.

Sl. no.	*Process*	*Machine code*	*Existing location*	*Existing total movement (m)*	*Optimum location*	*Optimum total movement (m)*
1	Cutting	M1	L_1	3.9	L_8	10.3
2	Centring/Facing	M2	L_2	0.5	L_9	0.5
3	Cross hole drilling	M3	L_3	0.5	L_1	14.83

Table 2.11: *Contd...*

Table 2.11: *Contd...*

Sl. no.	*Process*	*Machine code*	*Existing location*	*Existing total movement (m)*	*Optimum location*	*Optimum total movement (m)*
4	Finish turning	M4	L_4	0.5	L_2	0.5
5	To dispatch		L_{14}	11.1	L_{14}	6.3
			Total (Z_1)	16.5	Total (Z_1)	32.43

Comparison of total movement of mid ship shaft material for existing layout and optimum layout is presented in Table 2.12.

Comparison of total movement of sleeve material for existing layout and optimum layout is presented in Table 2.13.

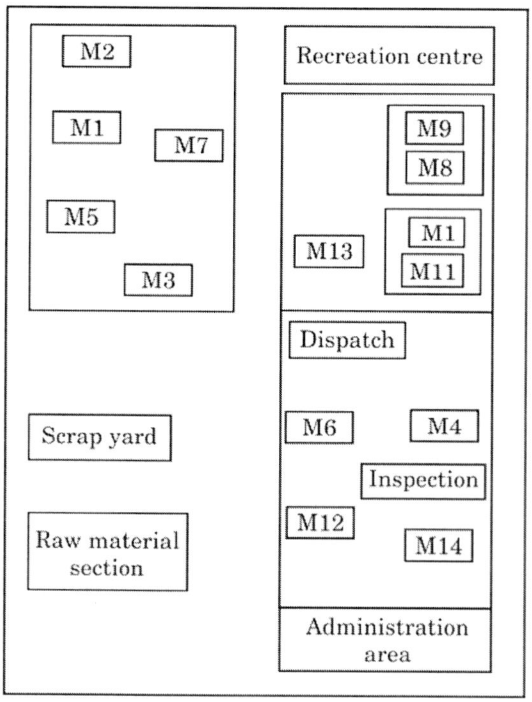

Fig. 2.4: Optimized plant layout in production of an automobile transmission system

Table 2.12: Total movement of mid ship shaft material for existing and optimum layout.

Sl. no.	*Process*	*Machine code*	*Existing location*	*Existing total movement (m)*	*Optimum location*	*Optimum total movement (m)*
1	Centring	M5	L_5	11.1	L_7	6.3
2	Centring	M5	L_5	0	L_7	0
3	Rough turning	M4	L_4	0.6	L_2	0.6
4	Rough turning	M6	L_6	1.1	L_3	0.5
5	Centre reaming	M5	L_5	0.5	L_7	1.1
6	Finish turning	M6	L_6	0.5	L_3	1.1
7	To dispatch		L_{14}	8.7	L_{14}	8.7
			Total (Z_2)	22.5	Total (Z_2)	18.30

Table 2.13: Total movement of sleeve material for existing and optimum layout.

Sl. no.	*Process*	*Machine code*	*Existing location*	*Existing total movement (m)*	*Optimum location*	*Optimum total movement (m)*
1	Cutting	M1	L_1	3.9	L_8	10.3
2	Centring	M7	L_7	3.02	L_{10}	0.4
3	Turning	M8	L_8	14.84	L_{13}	5.32
4	First side drilling	M10	L_{10}	0.4	L_{12}	0.5
5	Second side drilling	M9	L_9	3.01	L_{11}	0.5
6	Hardening	M14	L_{14}	13.33	L_{14}	14.13
7	Finish turning	M6	L_6	8.7	L_3	8.70
8	Second side boring	M11	L_{11}	20.55	L_5	1.10
9	First side boring	M12	L_{12}	0.5	L_4	0.6
10	Finish total length	M6	L_6	22.96	L_3	0.5
11	Drilling	M13	L_{13}	25.37	L_6	0.6
12	To dispatch		L_{14}	18.94	L_{14}	8.70
			Total (Z_3)	135.52	Total (Z_3)	51.35

The total material movement (Z) considering material movement of all three components is shown in Table 2.14.

It can be thus observed that the optimum layout obtained by RCGA provides significant improvement in the material movement of about 41.5% which will subsequently reduce the production time and operator fatigue as well.

Table 2.14: Total movement for existing layout and optimum layout.

Material movement for existing layout (m)				*Material movement for optimum layout (m)*			
Z_1	Z_2	Z_3	Z	Z_1	Z_2	Z_3	Z
16.5	22.5	135.52	174.52	32.43	18.30	51.35	102.08

2.4. Modeling and Optimization of Blank Nesting in Press Tool Operations

Material economy and resultant reduction in weight and cost, high productivity, use of unskilled labor, and a high degree of possible precision has rendered the presswork indispensable for many mass-produced goods. Hence, press tools contribute a major share in metal forming industries as steel metal stampings have now replaced many components, which are earlier cast or machined. However, due to continuous modification in product design, press tools are required to be designed frequently. The lead-time of manual design is very high, whereas due to emergence of modern computer controlled Electric Discharge Machine (EDM) and Rapid Prototyping (RP) techniques, manufacturing time is drastically reduced. To cope with the speed of manufacturing it is very essential to automate the design process. As a part of this, optimized blank nesting using evolutionary computational techniques has proved to be one of the best ways to minimize the design cycle time and to reduce material waste due to ineffective nesting of blanks. The term blank nesting is defined as getting the arrangement of a series of part dimensions on sheet metal so as to reduce the amount of metal scrap produced during the press operation as shown in Fig. 2.5 below.

As can be seen from Fig. 2.6, blank nesting is a function of blank orientation. If the component shown in Fig. 2.6 is arranged at 0° then utilization of strip is 100%, if arranged at 25° then utilization of strip is 68.2%, and if arranged at 45° then utilization of strip is only 50%. It is also to be noted that, besides orientation, pitch of successive blanks also plays a major role in reducing scrap.

A vast number of different industries are regularly faced with this problem of cutting two-dimensional objects from a stock-sheet as the first stage of their manufacturing process. Application areas explored in the literature include the metal stampings, garment industries, ship-building industries, cutting boards in the lumber, flat glass and furniture industries, cutting leather hides for footwear etc. In the case of metal stamping the requirement is often for large numbers of

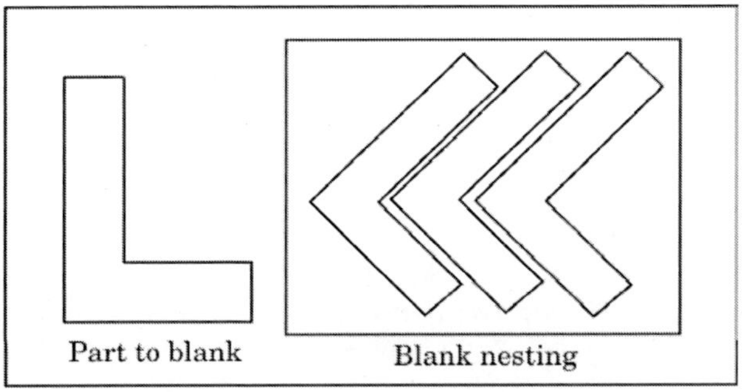

Fig. 2.5: Blank nesting configuration.

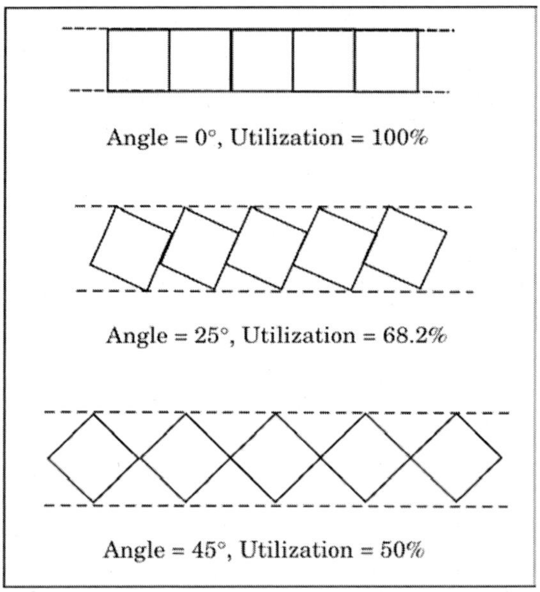

Fig. 2.6: Nesting configuration for a given blank orientation.

identical complex shapes however, orientation is not usually a restriction. The problem is basically to find an efficient way of nesting a given (small) number of copies of the piece together and then repeating this pattern along the stock-sheet. In the lumber cutting industry orientation may be important due to the grain of the wood. There is also a problem caused by imperfect or unusable areas on the stock-sheet. The leather industry faces these constraints along with complexities, as the stock-sheets are in fact individual hides which are themselves irregular. The parts nesting problem in the ship-building industry involves determining a cutting pattern for a given set of parts, each of which consists of a number of different pieces, from a series of metal stock-sheets. In spite of these differences all these problems have the common requirement of finding a feasible layout of the pieces on the stock-sheet.

The complexity of nesting problem thus depends on:

a) ***Dimensionality of the problem*:** Nesting of 3D parts is more complex than that of 2D parts.

b) ***Blank geometry***.

c) ***The number of shapes to be nested*:** Complexity increases with more number of parts to be nested. As shown in Fig. 2.7 for single shape to be nested only one orientation (θ_1) is to be considered. Whereas for two different shapes to be nested, two different orientations (θ_1 and θ_2) are to be considered. The number of orientations and thereby number of parameters to be optimized increases with number of shapes to be nested.

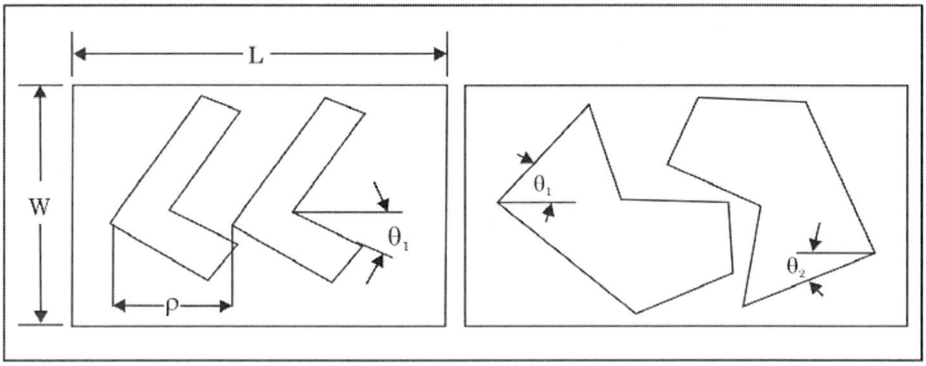

Fig. 2.7 : Complexity of problem with Number of shapes to be nested

d) ***Number of passes*:** The multi-pass nesting is shown in Fig. 2.8. As the number of passes increases (multi-pass design) the complexity of nesting problem is also increased

The problem considered in this case study, is blank nesting of curvilinear shapes in press tool operations. This problem can be formulated with an objective as to minimize the scrap (*i.e.,* to maximize the material utilization) by maximum possible squeezing of the blanks for a given orientation. However, while attempting to squeeze more for maximum utilization, an overlap between two blanks may occur making the layout infeasible. Hence, the constraint is to ensure that there should not be an overlap in optimized layout. The optimization model thus becomes:

$$\text{Minimize: Area of scrap } (S) = L \times W - n \times A_b. \tag{2.3}$$

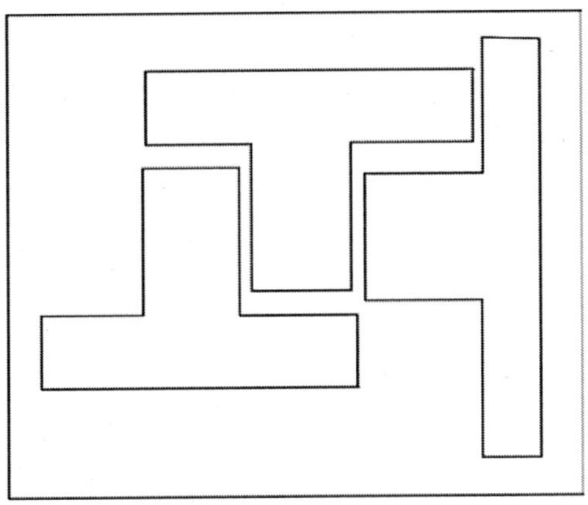

Fig. 2.8 : Complexity of problem with number of passes.

Where, L = length of strip, W = Width of strip for a given layout, n = maximum number of blanks cut from given sheet (= L/p, p is pitch for given layout) if scrap bridge is neglected, and A_b is area of blank.

Subject to: *Overlap area* (A_o) = 0 (2.4)

Variables: Blank Orientation (θ) and pitch (p).

Variable bounds: $0 \leq \theta \leq 360$; $p_{min} \leq p \leq p_{max}$

The overall objective function (z) is thus formulated as $z = S + penalty \times A_o$

The penalty value is to be so chosen that the layout with overlap should not appear as a final solution.

The overlap area 'A_o' is computed using following steps:

i) Identification of Point of Intersection

Intersection point is the point where one entity of root image gets intersected by the entity of neighborhood image. So these points are formed due to intersection of line and line, line and arc, arc and line, arc and arc as shown in Fig. 2.9.

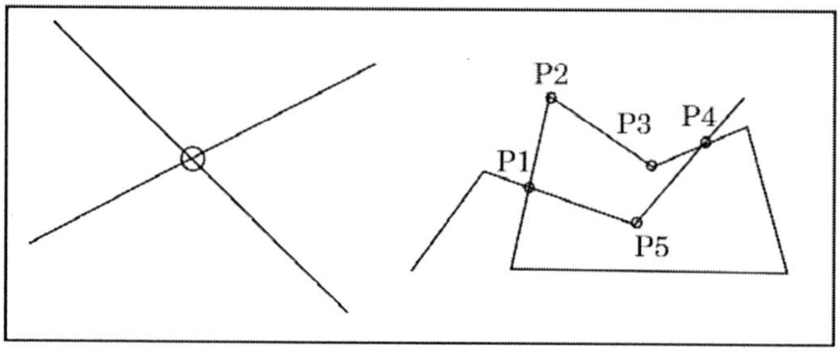

Fig. 2.9 : Identification of intersection points.

To calculate the intersection point of linear object, algorithm given by Avraham and Gary (1989) is implemented. For each edge of one polygon, find all the edges of the other polygon it intersects with and calculate all those intersection points. To calculate intersection of two lines, equations of lines are solved and intersection point is obtained. To determine the intersection of line and arc, line equation $y = m.x + c$ and circular arc equation $(x-h)^2 + (y-k)^2 = r^2$ are solved. The intersection point (x_{i1}, y_{i1}) or (x_{i2}, y_{i2}) of 2 arcs, having equations: $(x-a)^2 + (y-b)^2 = r^2$ and $(x-c)^2 + (y-d)^2 = m^2$ is calculated as follows:

Let, $e = c - a$ and $f = d - b$

Distance between centers of 2 arcs is

$$q = \sqrt{e^2 + f^2} \tag{2.5}$$

$$k = \frac{q^2 + r^2 - m^2}{2q} \tag{2.6}$$

$$X_{i1} = a + \frac{ek}{q} + \frac{f}{q}\sqrt{r^2 - k^2} \tag{2.7}$$

$$Y_{i1} = b + \frac{fk}{q} - \frac{e}{q}\sqrt{r^2 - k^2} \tag{2.8}$$

Or

$$X_{i2} = a + \frac{ek}{q} - \frac{f}{q}\sqrt{r^2 - k^2} \tag{2.9}$$

$$Y_{i2} = b + \frac{fk}{q} + \frac{e}{q}\sqrt{r^2 - k^2} \tag{2.10}$$

In Fig. 2.9, the points of intersection are P_1, P_4.

ii) Identification of Boundary Line Segment

Once the point of intersection is identified, then it is necessary to find the portion of the line, which forms the boundary of overlap region. This can be done by using algorithm given by Jain *et al.* (1992). The basic idea is to take the point say 'x_i' at an infinitesimal distance from the intersection point Q_i of two lines along the forward direction on one of the line (k to k+1) and compute the cross product of vectors (v and w) from x_i to end points of the other line (j to j+1) as shown in Fig. 2.10. The sign of the cross product of two vectors determines whether the segment is overlap boundary by indicating whether 'x_i' lies interior of other body.

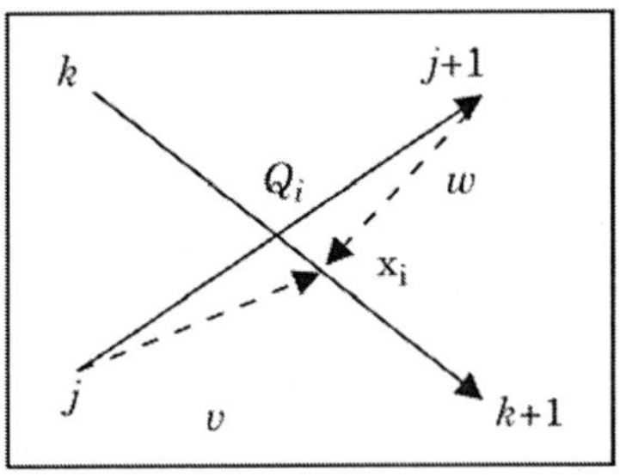

Fig. 2.10: Identification of boundary line segment (Jain *et al.*, 1992).

iii) *Identification of Overlap Region*

To find the overlap region, the line segment forming the boundary of overlap region (*viz* P1 to P5 in Fig. 2.11) is either connected to next point of intersection or end point of line on any object. Fig. shows P1–P2–P3–P4–P5 as overlap region. There may be more than one overlapping regions. The procedure is started at new point of intersection in that region that has not been traversed.

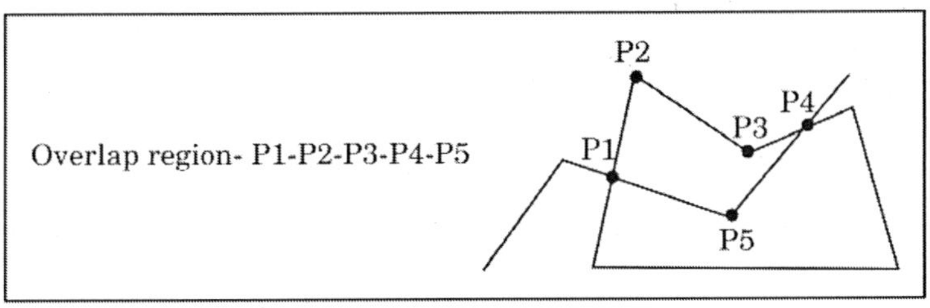

Fig. 2.11: Identification of overlap region.

iv) Calculating Area of Overlap Region

To find out the area of identified overlap region shown in Fig. 2.12, Green's theorem can be used. According to this theorem, for a closed boundary described by line segment with end points (x_i, y_i), overlap area is calculated using Eq.. (2.11).

$$A_{overlap} = \frac{1}{2}\sum x_i(y_{i+1} - y_i) - y_i(x_{i+1} - x_i) \tag{2.11}$$

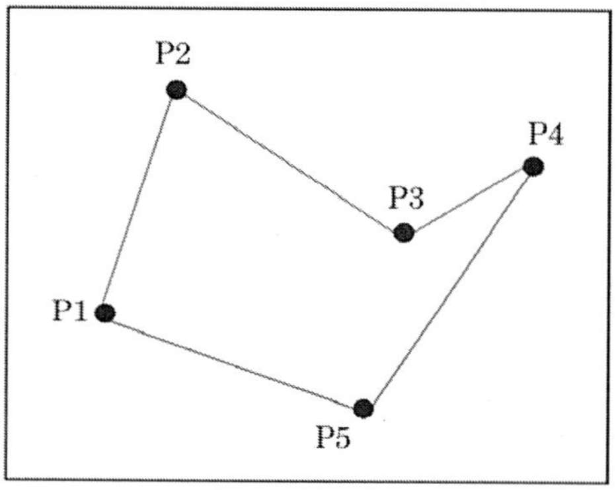

Fig. 2.12 : Calculating area of overlap region.

As drawing may also consist of circular arcs, Green's theorem calculates the area which covers the end point of the arcs as shown in Fig. 2.13.

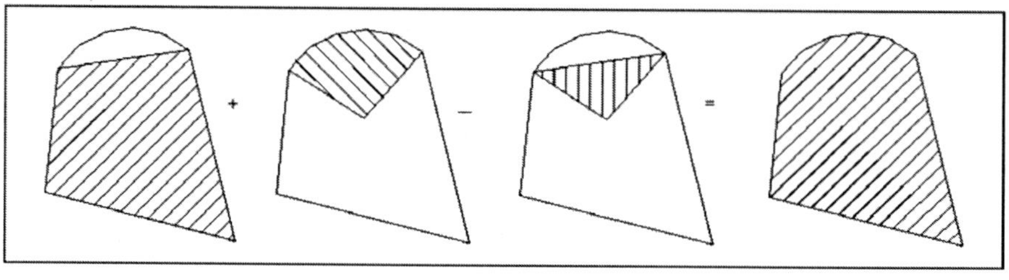

Fig. 2.13 : Area calculation of curvilinear shapes using Green's Theorem.

For arc sector, point inside or outside test is applied which determines arc is convex or concave. Area is calculated for arc sector and it is added or subtracted in the area calculated using Green's theorem.

Now the genetic algorithm is applied to determine the best layout for blank nesting for the component shown in Fig. 2.14 to be cut from a sheet of length (L) 250 cm. Width of sheet (W) will be layout generated. Variables are blank Orientation (θ) and pitch (p). Variable bounds: $0 \leq \theta \leq 2\pi$ (radian); $1 \leq p \leq 10$ cm. Penalty value is computed as 400.

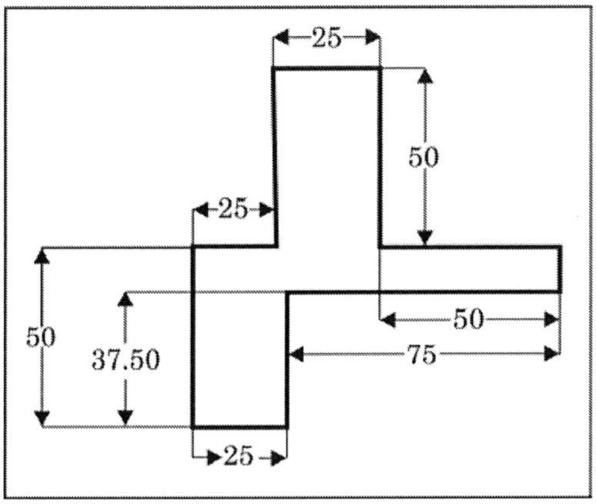

Fig. 2.14 : Component for blank nesting.

Step 1: *Generate randomly initial population of solution*

As genetic algorithm is population based algorithm, it starts with number of initial solutions. 10 initials solutions are considered as shown in Table 2.15.

Table 2.15: Random generation of initial population of solutions.

	Substring (x_1 for θ)		***Substring (x_2 for p)***				
Sl. no.	***Coded value***	***Decoded value***	***Coded value***	***Decoded value***	θ	p	z
1	1110	14	0110	6	5.86	4.60	2258.56
2	0101	5	1001	9	2.09	6.40	4270.42
3	1010	10	1010	10	4.19	7.00	1750.5
4	0001	1	1101	13	0.42	8.80	1813.58
5	1011	11	1000	8	4.60	5.80	1613.53
6	1100	12	1100	12	5.02	8.20	3592.63
7	1001	9	0100	4	3.77	3.40	1475.79
8	1111	15	0011	3	6.28	2.80	3031.25
9	1000	8	0000	0	3.35	1.00	3272.5
10	0111	7	0111	7	2.93	5.20	2204.59

Sample calculation of objective function value (z) for solution 1,:

Area of scrap (S) is computed as $(L\times W) - n\times A_b = 250\times9.13 - 54 \times 34.37 = 414.56$ cm^2

Where, Length of strip (L) = 250 cm, width of the layout (W) for θ = 5.86 rad is 9.13 cm as shown in Fig. 2.15.

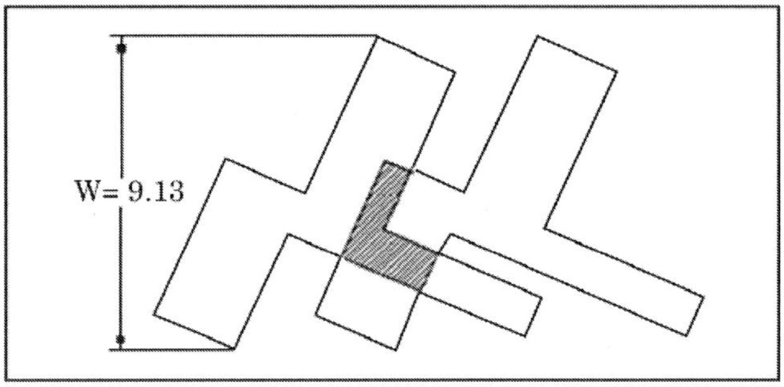

Fig. 2.15: Width of layout (W) and overlap area (A_o) for solution 1.

The number of blank produced (n) = L/p = 54 and area of each blank is 34.37 cm^2

Overlap area (A_o) for a given layout is shown with hatching lines in Fig. 2.15 = 4.61 cm^2.

Now, the objective function is computed as:

$$z = S + penalty \times A_o = 414.56 + 400 \times 4.61 = 2258.56$$

Step 2: *Reproduction*

The population of solutions at the end of reproduction is shown in Table 2.16.

Step 3: *Crossover and mutation*

The updated population of solutions after implementing crossover and mutation is presented in Table 2.17. Four pairs of solutions (6, 1), (8, 7), (3, 7) and (1, 9) undergo crossover. The bits marked with red are candidates for mutation.

Step 4: *Optimum solution*

The results of optimization are shown in Table 2.18. Fig. 2.16 shows optimum blank nesting configuration.

Table 2.16: Reproduction

Sl. no.	θ	*p*	*z*	*Fitness value*	*Individual strength*	*Relatitive strength*	*Cumu. relative strength*	*Rand. no.*	*Copied string* θ	*Copied string p*	*z*
1	5.86	4.60	2258.56	0.000443	0.992733	0.09927	0.09927	0.5425	5.02	8.20	3592.63
2	2.09	6.40	4270.42	0.000234	0.525042	0.05250	0.15178	0.0866	5.86	4.60	2258.56
3	4.19	7.00	1750.5	0.000571	1.280864	0.12809	0.27986	0.7268	3.77	3.40	1475.79
4	0.42	8.80	1813.58	0.000551	1.236313	0.12363	0.40350	0.7729	6.28	2.80	3031.25
5	4.60	5.80	1613.53	0.00062	1.389591	0.13896	0.54245	0.8274	6.28	2.80	3031.25
6	5.02	8.20	3592.63	0.000278	0.624097	0.06241	0.60486	0.6098	3.77	3.40	1475.79
7	3.77	3.40	1475.79	0.000678	1.519285	0.15193	0.75679	0.2296	4.19	7.00	1750.5
8	6.28	2.80	3031.25	0.00033	0.739679	0.07397	0.83076	0.6732	3.77	3.40	1475.79
9	3.35	1.00	3272.5	0.000306	0.68515	0.06851	0.89928	0.0687	5.86	4.60	2258.56
10	2.93	5.20	2204.59	0.000454	1.017035	0.10170	1.00098	0.8530	3.35	1.00	3272.5

Table 2.17: Crossover and mutation.

Sl. no.	*Reproduction*			*Crossover*		*Mutation*		*Solution (after first generation)*		
	x_1	x_2	*Site*	x_1	x_2	x_1	x_2	θ	*p*	*z*
6	1100	1100	3	1100	0110	1110	0110	5.86	4.60	2259.815
1	1110	0110		1110	1100	1110	1100	5.86	8.20	2483.884
7	1001	0100		1001	0100	0001	0100	0.42	3.40	1470.794
8	1111	0011		1111	0011	1111	1011	6.28	7.60	2569.408
8	1111	0011	2	1101	0100	1101	0100	5.44	3.40	3902.294
7	1001	0100		1011	0011	1011	0010	4.60	2.20	5061.318
3	1010	1010	5	1010	1100	1000	1100	3.35	8.20	1657.134
7	1001	0100		1001	0010	1001	0011	3.77	2.80	1562.25
1	1110	0110	6	1110	0100	0110	0000	2.51	1.00	2504.5
9	1000	0000		1000	0010	1000	1010	3.35	7.00	1801.5

Table 2.18: Parameter values for optimum blank nesting.

θ *(rad)*	θ *(°)*	*p (cm)*	*W (cm)*	*S* (cm^2)	A_o	*Z*
2.32	132.75	7.49	9.75 mm	1290	0	1290

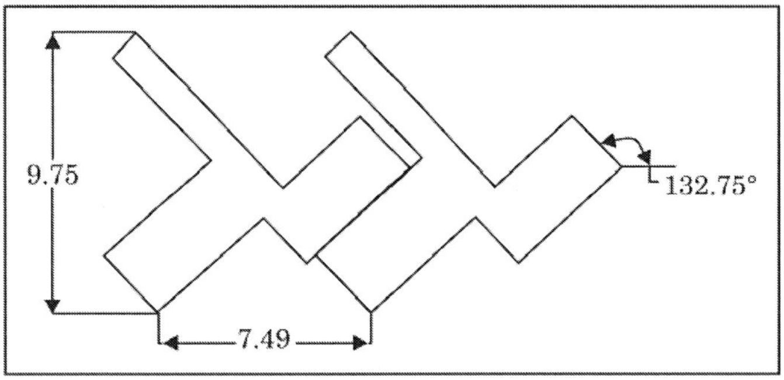

Fig. 2.16 : Optimum blank nesting configuration.

2.5. Material Flow Optimisation in Flexible Manufacturing System

Flexible manufacturing system (FMS) consists of integration of Computerized Numerical Controlled (CNC) machines for manufacturing of different parts and automated material handling system for conveying material from machine to machine and to the storage area with a central computer system for storing data and information and controlling the activities of machines and material handling systems. Customers are more concerned about the product quality and the demand of product is also changing drastically. So, to cater the needs of customers, it is inevitable for the manufacturing industries to enhance the flexibility and efficiency of the manufacturing system. FMS have two most important flexibilities *i.e.,* machine flexibility and routing flexibility. Machine flexibility means to manufacture different types of products on a machine whereas routing flexibility means manufacturing the operations on different machines. FMS is capable of achieving flexibility of low volume production as well as efficiency of high volume mass production (Gamila and Motavalli, 2003). Therefore, FMS development is a milestone in manufacturing industries. The investment required in FMS is high and so, a less payback period is desired by the manufacturing industries. This can be achieved by proper planning and execution in the FMS.

A case study of optimum selection of process plan and allocation of operations to machine so as to minimize the total cost of FMS is discussed in this section. The objective function formulated below is the extension of mathematical model developed by Paulo *et al.* (2002) where only three cost components namely operation cost, set up cost and material handling cost are considered. Bhosale and Pawar (2018) extended this model considering cost due to machine waiting also as given by Eq. (2.12). There is a set of '*n*' part types labelled with the indices $i = 1, \ldots n$; where the part type i has the known and uniform demand d_i over the planning period. A part type i can be processed under the different process plans $p = 1, \ldots P(i)$. For a process plan $p \in P(i)$ of a part type i, the manufacturing operations are represented by the indices $s=1, \ldots, S(ip)$. There is a set of m machines labelled with the indices $j= 1, \ldots, m$. The set of machines that can perform manufacturing operation s of part type i under process plan p is given by J_{ips}. The 0-1 decision variables are denoted by $X_{sj}(ip)$, where $X_{sj}(ip) = 1$ if operation s of (ip) is performed on machine j and zero otherwise.

$$X_{sj}(ip)=\begin{cases} 1 & \textit{if operation s is processed on machine j for the combination } (ip) \\ 0 & \textit{otherwise} \end{cases}$$

$OC_{sj}(ip)$ is the known per-unit cost of operation s of (ip) on machine j. This includes both the manufacturing and the re-fixturing costs. SC_j is the known set-up cost for machine j and the auxiliary variable M_j takes the value of one if machine j is selected and zero otherwise. $Ti(j+1)j$ is the cost of moving a unit of part type i from machine j to machine $j+1$ for the next operation. The waiting time is W_T.

$$C_T=\sum_{i=1}^{n} di \sum_{p=1}^{P(i)} \sum_{s=1}^{S(ip)} \sum_{j \in J_{ips}} OC_{sj}(ip) X_{sj}(ip) +$$

$$\sum_{j=1}^{m} SC_j M_j + \sum_{i=1}^{n} di \sum_{p=1}^{P(i)} \sum_{s=1}^{S(ip)} \sum_{j \in J_{ip(s+1)}} T_{ij(j+1)} L_{sj(j+1)}(ip) + \tag{2.12}$$

$$\sum_{i=1}^{n} di \sum_{p=1}^{P(i)} \sum_{s=1}^{S(ip)} \sum_{j \in J_{ip(s+1)}} W_{Ti(j+1)j} X_{sj}(ip) X_{(s+1)(j+1)}(ip)$$

As shown in Eq. (2.12) the first component is operation cost, second cost component is set up cost, third component is material handling cost and fourth component is cost due machine waiting.

The constraints are:

i) The first constraint is to ensure that each part type is processed under a single process plan, and it is given by Eq. (2.13):

$$\sum_{p=1}^{P(i)} Z(ip) = 1 \tag{2.13}$$

Where $Z(ip)$ = 1 if part type i is processed under process plan p and zero otherwise.

ii) The second constraint is to ensure that once a process plan is selected for a part type, each operation in that plan is processed on only one of the available machines and it is given by Eq. (2.14):

$$\sum_{j \in J_{ips}} X_{sj}(ip) = Z(ip) \quad \forall \text{i, p, s.} \tag{2.14}$$

The third constraint is to ensure that a machine j is selected before we assign any operations to it. It is given by Eq. (2.15):

$$\sum_{i=1}^{n} \sum_{p=1}^{P(i)} \sum_{s=1}^{S(ip)} X_{sj}(ip) \geq M_j \forall_j \tag{2.15}$$

iii) The fourth constraint is to ensure that the total time required by the operations allocated to a machine j, once it is selected, does not exceed the machine's known capacity. It is given by Eq. (2.16):

$$\sum_{i=1}^{n} \sum_{p=1}^{P(i)} \sum_{s=1}^{S(ip)} t_{sj}(ip) X_{sj}(ip) \leq b_j M_j \forall_j \tag{2.16}$$

Where t_{sj} (ip) is the time to perform operation s of (ip) on machine j. The two sets of constraints, (2.14) and (2.15), ensure consistency between the allocation of operations of the part types to machines and the selection of machines.

Now to demonstrate the application of real coded genetic algorithm, an application example is considered with 14 parts (i) which can be processed on 10 different machines. Available alternate process plans (p), the sequence of operations to be performed for selected process plan (s), the data for operation time and operation cost for this case study is presented in Table 2.19.

Table 2.19: Data for machines 1 to 5 (Paulo *et al.*, 2002).

			Machines									
i	***p***	***s***	***1***		***2***		***3***		***4***		***5***	
1	1	1	6	\$9			11	\$8	5	\$10		
		2	12	\$18							11	\$11
		3					12	\$9	10	\$20		
	2	1			6	\$8	9	\$7				
		2	4	\$6					13	\$26	11	\$11
	1	1	6	\$9		8	\$6	13	\$26			
		2							11	\$22	11	\$11
		1			7	\$9						
2	2	2							10	\$20	8	\$8
		3	4	\$6							12	\$12
	3	1			4	\$5	8	\$6				
	1	1							2	\$4	11	\$11
		2	11	\$17			12	\$9			7	\$7
3		1			7	\$9	4	\$3	8	\$16		
	2	2	2	\$3					11	\$22		
		3							2	\$4	11	\$11
		4	10	\$15			10	\$8				
	1	1							10	\$20	7	\$7
		2	7	\$11					8	\$16		
4		1		9	\$11						9	\$9
	2	2	6	\$9			12	\$9	2	\$4		
		3		11	\$14						11	\$11
		1	4	\$6							13	\$13
	1	2			2	\$3	4	\$3				
		3	13	\$20	13	\$16	9	\$7				
5		1					6	\$5	2	\$4		
	2	2	2	\$3	8	\$10						
	3	1										
		2	2	\$3					13	\$26	7	\$7
6	1	1	5	\$8	6	\$8	10	\$8				
		2			7	\$9						
		3							4	\$8		
		4	3	\$5	3	\$4					7	\$7

Table 2.19: (*Contd...*)

Table 2.19: (*Contd...*)

			Machines									
i	*p*	*s*	*1*		*2*		*3*		*4*			
7	1	1	10	$15								
		2					7	$5				
	2	1			11	$14						
		2	7	$11					9	$18		
		1			2	$3						
8	1	2	4	$6					11	$22	2	$2
		3	7	$11			5	$4				
	2	1					7	$5				
		2	6	$9								
9	1	1					4	$3				
		2	11	$17	12	$15						
	1	1									13	$13
		2	2	$3	12	$15	9	$7				
		1	3	$5							5	$5
10	2	2					7	$5	11	$22		
		3	4	$6					2	$4	8	$8
	3	1					10	$8	12	$24		
		2	8	$12							3	$3
	1	1	4	$6							3	$3
		2					4	$3	11	$22		
11		1			11	$14						
	2	2	12	$18			9	$7	13	$26		
		3	9	$14							9	$9
		1			7	$9	8	$6				
	1	2	4	$6					6	$12	13	$13
12		3			5	$6						
		1	10	$15					4	$8		
	2	2					6	$5				
13	1	1									8	$8
		2			12	$15			11	$22	3	$3
		3	3	$5								
		4					7	$5			5	$5
14	1	1			3	$4					10	$10
		2	10	$15							13	$13
	2	1			13	$16	2	$2				
		2	4	$6							9	$9

Table 2.19: (*Contd...*)

Table 2.19: *(Contd...)*

			Machines				
i	***p***	***s***	***6***	***7***	***8***	***9***	***10***
1	1	1		11 $7			7 $8
		2		5 $3	3 $3		
		3					8 $10
	2	1	7 $10				
		2	6 $8				8 $10
2	1	1				8 $13	
		2	5 $7		2 $2		
	2	1		8 $5	8 $7	13 $21	7 $8
		2			2 $2	11 $18	
		3				7 $11	6 $7
	3	1		8 $5			4 $5
		2	10 $14			9 $14	
3	1	1				2 $3	13 $16
		2		4 $2		9 $14	
	2	1		13 $8			
		2	2 $3				
		3			13 $12		4 $5
		4	9 $13	10 $6			2 $2
4	1	1				7 $11	
		2	9 $13	3 $2			
	2	1		8 $5	9 $8		3 $4
		2				6 $10	13 $16
		3	5 $7		5 $5	4 $6	
5	1	1					5 $6
		2			8 $7	3 $5	
		3					10 $12
	2	1			5 $5		
		2					5 $6
	3	1	12 $17	4 $2	8 $7		
		2			11 $10		
6	1	1					12 $14
		2	13 $18	3 $2			
		3		5 $3	10 $9	11 $18	6 $7
		4		7 $4			
7	1	1	8 $11	5 $3		9 $14	9 $11
		2		10 $6	11 $10		
	2	1				11 $18	9 $11
		2			8 $7		
8	1	1	9 $13	7 $4			2 $2
		2			7 $6	4 $6	
		3		5 $3			

Table 2.19: *(Contd...)*

Table 2.19: (*Contd...*)

			Machines				
i	***p***	***s***	***6***	***7***	***8***	***9***	***10***
	2	1	10 $14				
		2		8 $5			11 $13
9	1	1	7 $10			12 $19	7 $8
		2				10 $16	11 $13
10	1	1		3 $2	11 $10		
		2	3 $4				
	2	1			13 $12	9 $14	9 $11
		2	9 $13				7 $8
		3			6 $5		
	3	1			11 $10		
		2	11 $15				4 $5
11	1	1	10 $14	11 $7			
		2		10 $6		8 $13	
	2	1	9 $13			7 $11	11 $13
		2					
		3				11 $18	6 $7
12	1	1					12 $14
		2	4 $6				
		3		4 $2		9 $14	2 $2
	2	1	3 $4				
		2		6 $4		10 $16	13 $16
13	1	1	9 $13		4 $4		
		2				11 $18	
		3			11 $10	5 $8	3 $4
		4					4 $5
14	1	1	5 $7	10 $6			
		2			2 $2		
	2	1		10 $6	4 $4	12 $19	
		2	3 $4				

Table 2.19 shows that, all products have one or more process plans and each process plan has one or more operations. It is assumed that the details regarding number of jobs to be produced, their processing time, number of operations to be performed on each product, are already available in advance. In this problem, the flexibility of selecting a process plan and then allocating an operation to a machine is available. This flexibility makes this problem complex in nature due to selection of best alternative out of large number of possible solutions. Considering '*j*' as total number of products having 'o_j' operations to be processed on '*m*' machines, the number of possible solutions can be calculated using Eq. (2.17) (Abazari *et al.,* 2012):

$$\text{Number of Possible solutions} = \prod_{j=1}^{j}\left\{\left(\prod_{o=1}^{o_j} m_{o_j}\right)+1\right\} \quad (2.17)$$

The illustration for the example given in Table 2.19, is as follows:

$$\prod_{j=1}^{14}\left\{\left(\prod_{o=1}^{4} m_{o_j}\right)+1\right\} =$$

((5×4×3)+1) × ((3×5)+1) × ((4×4)+1) × ((5×4×4)+1) × ((4×3)+1) × ((4×5)+1) × ((4×3×4×5)+1) × ((3×4)+1) × ((5×5×5)+1) × ((3×4×4)+1) × ((3×3)+1) × ((3×4)+1) × ((4×3×5×4)+1) × ((5×3)+1) × ((3×3)+1) × ((4×5×3)+1) × ((2×3)+1) × ((4×4)+1) × ((5×4×3)+1) × ((3×5)+1) × ((4×4)+1) × ((5×4×4)+1) × ((4×3)+1) × ((4×5)+1) × ((4×3×4×5)+1) × ((3×4)+1) × ((5×5×5)+1) = 3.7×10^{40}

Now, to find the best solution out of 3.7×10^{40} satisfying all constraints is a challenging work and hence real coded genetic algorithm is applied for the present case.

The machine set up cost (*SC*) for different machines (machine number mentioned in bracket) in $ is as follows: SC(1) = 120, SC(2) = 230, SC(3) = 450, SC(4) = 60, SC(5) = 180, SC(6) = 220, SC(2) = 310, SC(8) = 90, SC(9) = 260, SC(10) = 550.

Table 2.20 below shows the cost of moving a part from one machine to another machine.

Steps of RCGA implemented to solve the above model resulted into optimal solution shown in Table 2.21.

The part which is having minimum processing time will be processed first and then the part having second minimum processing time will be processed next. Likewise, according to the processing time of the parts, they are selected and processed on different machines. The product 1 is having two operations and first operation with processing time of 180 time units will be performed on machine 2 and then second operation having processing time of 120 time units will be performed on machine 1. But, as the product 14 is having first operation of 120 time, it will be processed first (as it is minimum one) on machine 2. So, machine 2 will process first operation of product 14 for 120 time units and meanwhile operation 1 of product 1 will have waiting time of 120 time units. When the first operation on machine 2 is over then operation 1 of product 1 will be taken for processing, which is having processing time of 180 minutes. So, the completion time for operation 1 of product 1 is 120 + 180 = 300 time units. Thus the product 1 is ready in 300 time units for further operation 2. But, as machine 1 is busy in processing another operations and it will be finish operations at 460 time units, operation 2 of product 1 will have a total waiting of 460 time units. It has processing time of 120 time units, so, the completion time is 460 + 120 = 580 time units. Similarly, the completion time for each product on each machine is evaluated and is presented in Table 2.22.

Table 2.20: Cost ($) of moving parts from machine to machine (Paulo *et al.*, 2002).

Machine no.	*1*	*2*	*3*	*4*	*5*	*6*	*7*	*8*	*9*	*10*
1	0	4	14	11	5	9	13	10	11	7
2	4	0	12	13	7	7	6	13	5	4
3	14	12	0	7	10	4	5	11	4	9
4	11	13	7	0	13	13	6	14	12	7
5	5	7	10	13	0	5	15	6	10	15
6	9	7	4	13	5	0	6	11	9	4
7	13	6	5	6	15	6	0	8	5	6
8	11	13	11	14	6	11	8	0	5	15
9	11	5	4	12	10	9	5	5	0	9
10	7	4	9	7	15	4	6	15	9	0

Table 2.21: Results of material flow optimization using RCGA.

Part number	*1*	*2*	*3*	*4*	*5*	*6*	*7*	*8*	*9*	*10*	*11*	*12*	*13*	*14*
Process plan	2	1	1	1	2	1	2	1	1	1	1	2	1	1
Machine 1	2	1			2			3		2			3	
Machine 2	1					1, 4								1
Machine 3											2	2		
Machine 4				1	1	3								
Machine 5								2			1		2, 4	
Machine 6												1	1	
Machine 7				2		2				1				
Machine 8		2					2							2
Machine 9			1, 2						2					
Machine 10							1	1	1					

Table 2.22: Machine set up cost and material handling cost for optimum solution.

P	***Process Plan***		***Machines***					
			1		***2***	***3***	***4***	***5***
1	2 (2, 1)	*S*	300	0				
		O	120	180				
		W	460	120				
		F	580	300				
2	1(1, 8)	*S*	0					
		O	300					
		W	0					
		F	300					
3	1 (9, 9)	*S*						
		O						
		W						
		F						
4	1(4, 7)	*S*					0	
		O					650	
		W					50	
		F					700	
5	2(4, 1)	*S*	50				0	
		O	50				50	
		W	300				0	
		F	350				50	
6	1(2, 7, 4, 2)	*S*		0	860		660	
		O		240	120		160	
		W		300	540		700	
		F		540	980		860	
7	2(10, 8)	*S*						
		O						
		W						
		F						
8	1(10, 5, 1)	*S*	220					40
		O	140					40
		W	580					180
		F	720					220
9	1(10, 9)	*S*						
		O						
		W						
		F						
10	1(7, 1)	*S*	165					
		O	110					
		W	350					
		F	460					

Table 2.22: (*Contd...*)

Table 2.22: (*Contd...*)

P	*Process Plan*		*Machines*				
			1	*2*	*3*	*4*	*5*
11	1(5, 3)	*S*			180	0	
		O			240	180	
		W			315	0	
		F			555	180	
12	2 (6, 3)	*S*			105		
		O			210		
		W			0		
		F			315		
13	1(6, 5, 1, 5)	*S*	1065			825	1305
		O	240			240	400
		W	720			220	460
		F	1305			1065	1705
14	1(2, 8)	*S*		0			
		O		120			
		W		0			
		F		120			
M/C Set up Cost			120	230	450	60	180
Material handling Cost				149			

P = *Product, S* = Start time, *O* = Operation or processing time, *W* = Waiting time, *F* = Finish time.

Table 2.22: (*Contd...*)

P	*Process Plan*		*Machines*				
			6	*7*	*8*	*9*	*10*
1	2 (2, 1)	*S*					
		O					
		W					
		F					
2	1(1, 8)	*S*			300		
		O			100		
		W			200		
		F			400		
3	1 (9, 9)	*S*				0 90	
		O				90 405	
		W				0 295	
		F				90 700	
4	1(4, 7)	*S*		700			

Table 2.22: (*Contd...*)

Table 2.22: (*Contd...*)

P	*Process Plan*		*Machines*				
			6	***7***	***8***	***9***	***10***
		O		195			
		W		660			
		F		895			
5	2(4, 1)	*S*					
		O					
		W					
		F					
6	1(2, 7, 4, 2)	*S*		540			
		O		120			
		W		165			
		F		660			
7	2(10, 8)	*S*			955		0
		O			720		810
		W			400		145
		F			1675		955
8	1(10, 5, 1)	*S*					0
		O					40
		W					0
		F					40
9	1(10,9)	*S*				145	0
		O				150	105
		W				90	40
		F				295	145
10	1(7, 1)	*S*		0			
		O		165			
		W		0			
		F		165			
11	1(5, 3)	*S*					
		O					
		W					
		F					
12	2 (6, 3)	*S*	0				
		O	105				
		W	0				
		F	105				
13	1(6, 5, 1, 5)	*S*	0				
		O	720				
		W	105				
		F	825				

Table 2.22: (*Contd...*)

Table 2.22: (*Contd...*)

P	*Process Plan*		*Machines*				
			6	*7*	*8*	*9*	*10*
14	1(2, 8)	*S*			120		
		O			80		
		W			0		
		F			200		
M/C Set up Cost			220	310	90	260	550

The comparison of total cost for optimum solutions obtained by Bhosale and Pawar (2018), Paulo *et al.* (2002) and Sujono and Lashkari (2007) are presented in Table 2.23.

Table 2.23: Comparison of results of optimization for case study 2.3.

Sl. no.	*Sequence of operation*	*Operation cost ($)*	*Material handling ($)*	*Machine set up t cost ($)*	*Waiting cost ($)*	*Total cost ($)*
		a	*b*	*c*	*d*	*=a+b+c+d*
Bhosale and Pawar 2018)	1(2){2,1}*, 2(1){1,8}, 3(1){9,9}, 4(1){4,7}, 5(2){4,1}, 6(1){2,7,4,2}, 7(2){10,8}, 8(1){10,5,1}, 9(1){10,9}, 10(1){7,1}, 11(1){5,3}, 12(2){6,3}, 13(1){6,5,1,5}, 14(1){2,8}	9755	150	2470	175	12455
Paulo *et al.* (2002)	1(2){2,1}, 2(1){1,8}, 3(1){9,9}, 4(1){9,7}, 5(2){4,1}, 6(1){2,7,4,2}, 7(2){10,8}, 8(1){10,5,1}, 9(1){10,2}, 10(1){7,1}, 11(1){5,3}, 12(2){6,3}, 13(1){6,5,1,5}, 14(1){2,8}	9170	149	2470	1255	13044
Sujono and Lashkari (2007)	1(2){2,1}, 2(1){1,8}, 3(1){4,7}, 4(1){5,7}, 5(3){7,1}, 6(1){2,7,8,2}, 7(2){7,8}, 8(1){2,5,7}, 9(1){6,2}, 10(1){7,1}, 11(1){5,7}, 12(2){6,7}, 13(1){8,5,1,5}, 14(1){2,8}	7220	175	1210	6165	14820

* First part is processed by process plan 2 in which operation 1 is carried out on machine 2 and operation 2 is carried out on machine 1.

The results show that operation cost obtained by Paulo *et al.* (2002) and by Sujono and Lashkari (2007) is less as compared to that obtained by Bhosale and Pawar (2018). This is due to the fact that, Paulo *et al.* (2002) and Sujono and Lashkari (2007) had not considered waiting time cost in their model. However if waiting time cost is considered for the solutions obtained by Paulo *et al.* (2002) and by Sujono and Lashkari (2007), the total cost for the solution obtained by Bhosale and Pawar (2018) is less than that obtained by Paulo *et al.* (2002) as well as by Sujono and Lashkari (2007). Also, the solution obtained by Bhosale and Pawar (2018) also reveals the effectiveness of real coded genetic algorithm to solve this problem.

Although, genetic algorithm can be thus effectively applied to solve many real life problems, the performance of the genetic algorithm mainly depends on algorithm specific parameters such as population size, number of generations, crossover rate and mutation rate, etc. Genetic algorithms are preferred when near optimal improved conditions instead of exact optimum solution are cost effective and acceptable for implementation by the manufacturer. It is a derivative free approach for near optimal points search direction and may be applied to continuous or discrete response function. Although GA has advantages over the traditional techniques, it has following limitations:

i) In GA, all offspring are accepted and their parent strings are abandoned at the end of every generation regardless of their fitness values. This gives rise to a risk that a good parent string may be replaced with its deteriorated child string. Thus, the improvement on the average performance of child population over parent population cannot be always guaranteed.

ii) In GA, only good parent strings are given chance to produce offspring without any consideration of the possibilities of generating better offspring by others.

iii) They are not efficient when convergence speed is taken into consideration.

iv) Genetic algorithm does not guarantee optimal solution.

To demonstrate this fact, genetic algorithm toolbox of Matlab is used to solve optimization model for ultrasonic machining process discussed in section 1.2 of Chapter 1. The result of optimization for the first run (using default GA parameters) shows optimum value of Z_1 = 3.142 as shown in Fig. 2.17.

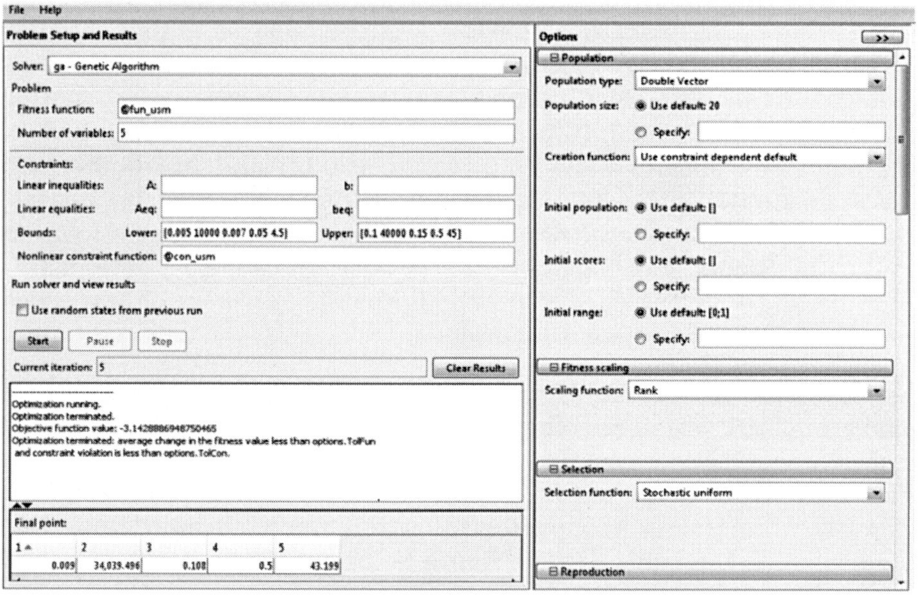

Fig. 2.17: Solution obtained by GA for problem in case study 1.1 (trial run 1).

The second run is carried out with another set of GA parameters which result into optimum value of Z_1 = 2.746 as shown in Fig. 2.18. Thus, it is observed that in both cases the genetic algorithm converges to local optimum solution. In fact the success rate of GA itself is low. This is evident from Fig. 2.19 which shows the comparison of various algorithms for solving standard benchmark function for continuous optimization known as Griewank function which is expressed by Eq. (2.18).

$$f(x) = 1 + \frac{1}{4000}\sum_{j=1}^{n} x_j^2 - \prod_{j=1}^{n} \cos\left(\frac{x_j}{\sqrt{j}}\right) \tag{2.18}$$

It is observed from Fig. 2.19, that the performance of GA is inferior to other algorithms when compared with respect to success rate. Similar is case for convergence rate. This is mainly due to the fact that besides other factors, GA requires large number of algorithm specific parameters to be tuned for specific application. Hence attempts have been made by researchers to develop several variants of genetic algorithm and also the algorithms which requires less number of parameters to be tuned.

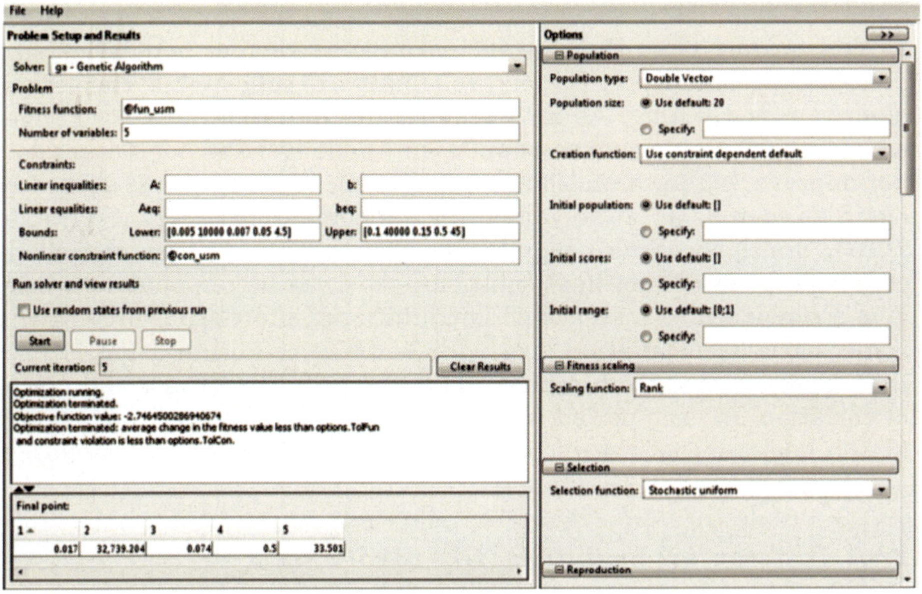

Fig. 2.18: Solution obtained by GA for problem in case study 1.1 (trial run 2).

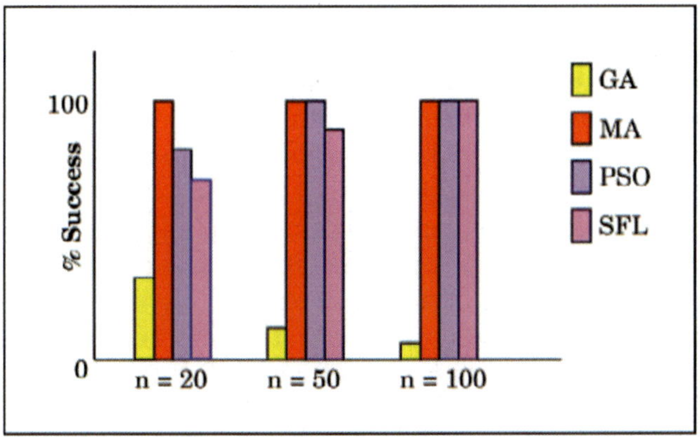

Fig. 2.19: Comparison of GA with other algorithms (Elbeltagi *et al.*, 2005).

2.6. Variants of Genetic Algorithm

A number of methods have been proposed to improve the robustness and computational efficiency of GAs. Few of them are discussed below.

2.6.1. *Adaptive genetic algorithm*

In this variant, algorithm specific parameters such as population size, crossover probability, mutation probability are tuned while the genetic algorithm is running. As an example, the mutation rate is changed according to changes in the population; the longer the population does not improve, higher mutation rate is chosen. *Vice versa*, it is decreased again as soon as an improvement of the population occurs (Sharapov, 2007). Similarly, Small population sizes may result in premature convergence to non-optimal solutions, whereas large population sizes give a considerable increase of computational effort. To overcome this issue, variable population size with periodic re-initialization is used that follows a saw-tooth scheme with a specific amplitude and period of variation. In each period, the population size decreases linearly and at the beginning of the next period randomly generated individuals are appended to the population.

2.6.2. *Differential Evolution (DE) algorithm*

The Differential Evolution (DE) algorithm proposed by Storn and Price (1997) is a population based algorithm like genetic algorithms using the similar operators; crossover, mutation and selection. The main difference in constructing better solutions is that genetic algorithms rely on crossover while DE relies on mutation operation. This operation is based on the differences of randomly sampled pairs of solutions in the population. The algorithm uses mutation operation as a search mechanism and selection operation to direct the search toward the prospective regions in the search space. The DE algorithm also uses a non-uniform crossover that can take child vector parameters from one parent more often than it does from others. By using the components of the existing population members to construct trial vectors, the recombination (crossover) operator efficiently shuffles information about successful combinations, enabling the search for a better solution space.

The Differential Evolution algorithm involves maintaining a population of candidate solutions subjected to iterations of recombination, evaluation, and selection. The recombination approach involves the creation of new candidate solution components based on the weighted difference between two randomly selected population members added to a third population member. This perturbs population members relative to the spread of the broader population. In conjunction with selection, the perturbation effect self-organizes the sampling of the problem space, bounding it to known areas of interest. The algorithm has following merits:

i) Able to find the true global minimum regardless of the initial parameter values.
ii) Fast convergence.
iii) Needs few control parameters to be tuned.
iv) Easily adaptable for integer and discrete optimization.
v) Quite effective in nonlinear constraint optimization including penalty functions.
vi) Useful for optimizing multi-modal search spaces.

2.6.3. *Memetic algorithm*

The genetic algorithm is not well suited for fine-tuning structures which are close to optimal solution. The memetic algorithm proposed by Moscato (1989) can be viewed as a marriage between a population-based global technique and a local search made by each of the individuals. It is a special kind of genetic algorithms with a local hill climbing. Like genetic algorithms, memetic algorithm is a population-based approach. It has shown that it is an order of magnitude faster than traditional genetic algorithms for some problem domains. In a memetic algorithm the population is initialized at random or using a heuristic. Then, each individual makes local search to improve its fitness. To form a new population for the next generation, higher quality individuals are selected. The selection phase is identical to that used in the classical genetic algorithm selection phase. Once two parents have been selected, their chromosomes are combined and the classical operators of crossover are applied to generate new individuals. The latter are enhanced using a local search technique. The role of local search in memetic algorithms is to locate the local optimum more efficiently then the genetic algorithm.

3

Particle Swarm Optimization

3.1. Introduction

Particle swarm optimization (PSO) is an evolutionary computational technique developed by Kennedy and Eberhart (1995). It belongs to the field of Swarm Intelligence and Collective Intelligence and is a sub-field of Computational Intelligence. Particles in the swarm fly through an environment following the fitter members of the swarm and generally biasing their movement toward historically good areas of their environment. It exhibits common evolutionary computation attributes including initialization with a population of random solutions (n_p) and searching for optima by updating through number of generations (n). Potential solutions, called particles, are then "flown" through the problem space by following the current optimum particles. The updates of the particles are accomplished as per the following equations.

$$V_{i+1} = w \times V_i + c_1 \times r_1 \times (pBest_i - X_i) + c_2 \times r_2 \times (gBest_i - X_i) \tag{3.1}$$

$$X_{i+1} = X_i + V_{i+1} \tag{3.2}$$

Eq. (3.1) calculates a new velocity (V_{i+1}) for each particle (potential solution) based on its previous velocity, the best location it has achieved ('*pBest*') so far, and the global best location ('*gBest*'), the population has achieved. Eq. (3.2) updates individual particle's position (X_i) in solution hyperspace. The two random numbers 'r_1' and 'r_2' in Eq. (3.1) are independently generated in the range [0, 1]. The acceleration constants 'c_1' and 'c_2' in Eq. (3.1) represent the weighting of the stochastic acceleration terms that pull each particle towards '*pBest*' and '*gBest*' positions. 'c_1' represents the confidence the particle has in itself (cognitive parameter) and 'c_2' represents the confidence the particle has in swarm (social parameter). Thus, adjustment of these constants changes the amount of tension in the system. The inertia weight '*w*' plays an important role in the PSO convergence behavior since it is employed to control the exploration abilities of the swarm.

To achieve the dimensional consistency of equations (3.1) and (3.2), the dimension of the term '$c \times r$' in Eq. (3.1) could be taken as $(\text{time})^{-2}$. This way, the second and the third terms in Eq. (3.1) assume the dimension of acceleration. To get the correct dimension of velocity, as required by the left-hand side, one needs to multiply them by Δt, the time step, which becomes unity in the present case, denoting changes from iteration i to $i+1$. Similarly, the second term in Eq. (3.2) assumes the correct dimension when taken as $V_{i+1}\Delta t$ and the present form results through the implicit assumption that $\Delta t = 1$. Particle's velocities on each dimension are confined to a maximum velocity parameter V_{max}, specified by the user. If the sum of accelerations would cause the velocity on that dimension to exceed V_{max}, then the velocity on that dimension is limited to V_{max}. As the genetic algorithm and particle swarm optimization both are evolutionary algorithms, they share some common features as given below:

- Both algorithms start with randomly generated population.
- Evaluates the population with fitness value.
- Updates the position and search with random techniques.
- Both algorithms do not guarantee optimal solution.

However, the PSO algorithm differs from genetic algorithm with respect to following points:

- Unlike genetic algorithm, PSO does not need complex encoding and decoding process and special genetic operator, it takes real number as a particle in the aspect of representation solution.
- PSO does not need genetic operator like cross over and mutation. The particles update themselves with internal velocity. (However, modified versions of PSO uses mutation operator to avoid the premature convergence).
- Unlike GA, all particles tend to converge to the best solution in case of PSO.

- In PSO, it is difficult to keep the diversity of the population. The search rates are relatively low and hence it may require more computation time while solving the complex optimization problem.

3.2. Mechanism of Working of Particle Swarm Optimization (PSO) Algorithm

As mentioned earlier, the two key operations in PSO are the update of velocity and the update of position. The velocity is updated based on three components: (1) the old velocity (inertia or momentum term), (2) experience of an individual particle (cognitive or self learning term), and (3) experience of the whole swarm (group or social learning term). Each term has a weight constant associated with it. For basic PSO algorithm, the number of required constants is three: inertia weight (w), cognitive parameter (c_1), social parameter (c_2). Particle swarm optimization does not require sorting of fitness values of solutions in any process. This reduces the computational efforts required as compared to genetic algorithm, especially when the population size is large. The updates of velocity and position in PSO also only require a simple arithmetic operation of real numbers. The flowchart of particle swarm optimization is shown in Fig. 3.1.

The working of particle swarm optimization algorithm is demonstrated through a simple example discussed below.

Consider an objective function:

$Minimize\ z = x_1^2 - 3x_2 + 10$ such that $0 \leq x_{1,}x_2 \leq 3$.

Steps of PSO algorithm

Step 1: ***Generate randomly initial population of solution***

Considering five birds in a swarm, Fig. 3.2 shows initial position of birds (x_1, x_2) and corresponding fitness values (f_i).

Step 2: ***Identify pBest and gBest***

As shown in Fig. 3.2, being initial position, for every bird the current position is its own best (*pBest*). Since the problem is of minimization, it can be seen that bird 1 is having highest fitness value hence is global best (*gBest*).

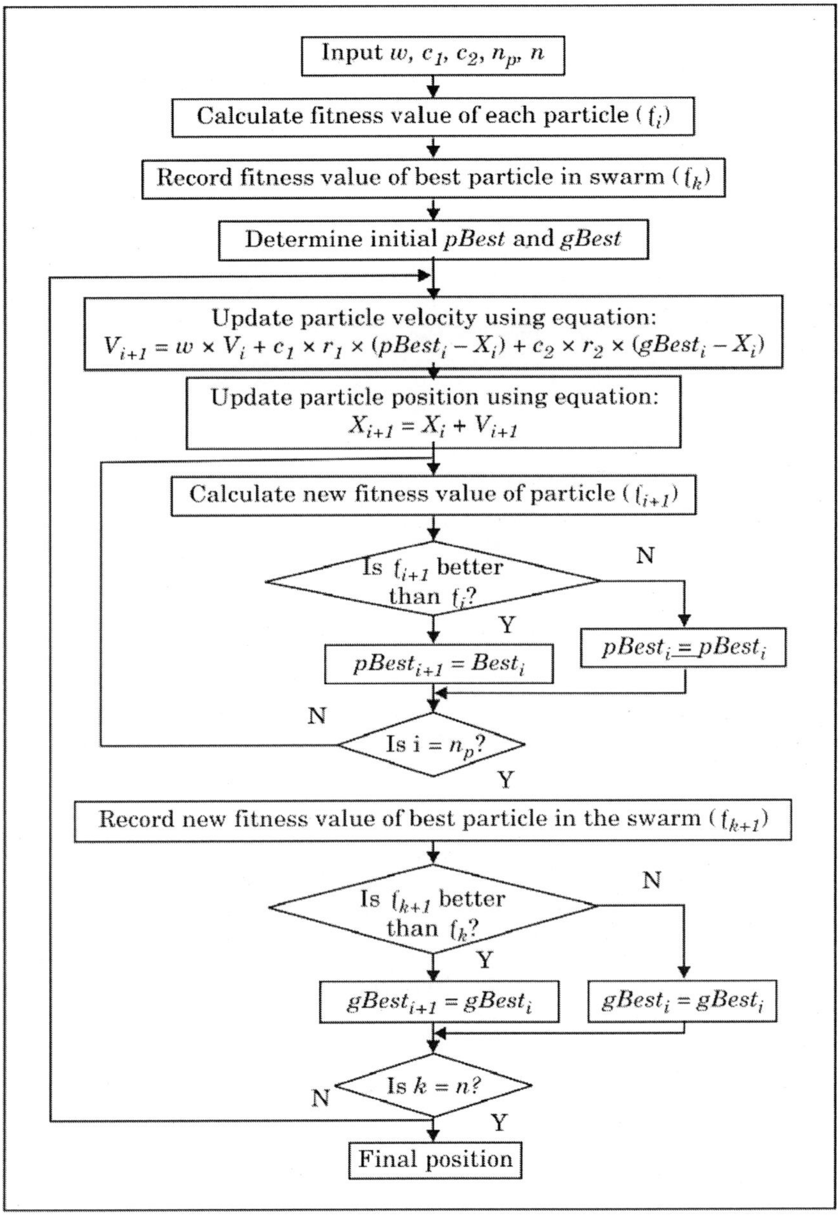

Fig. 3.1: Flowchart for particle swarm optimization (PSO) algorithm (Kennedy and Eberhart, 1995).

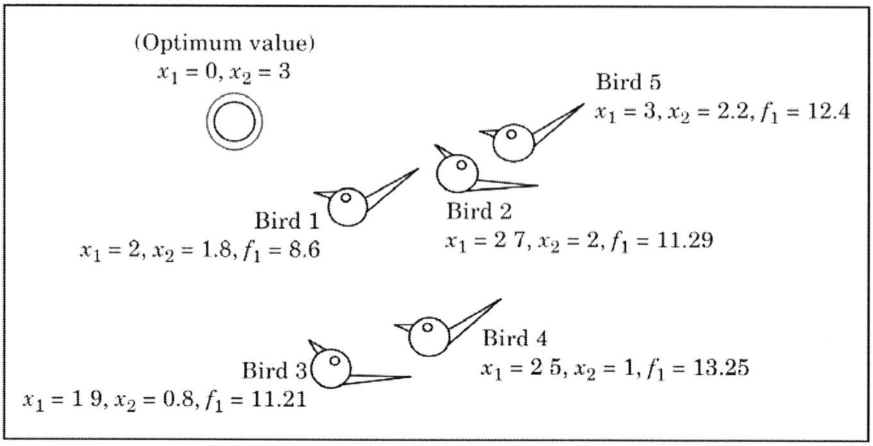

Fig. 3.2 : Initial position of birds (initial population of solutions).

Step 3: *Update positions of birds*

Let the initial velocities (v_1, v_2) be (0.43, 0.11), (1.36, 0.49), (0.01, 0.42), (0.33. 1.87), and (0.07, 1.03) for birds 1 to 5 respectively. Consider the values of three coefficients as: Inertia coefficient (w) = 0.65, cognitive parameter (c_1) = 1.65, and social parameter (c_2) = 1.75. The criteria for selection and tuning of these coefficients are discussed in section 3.3. Now, Eq. (3.1) and Eq. (3.2) are used to determine new velocities and new positions respectively. These new velocities are calculated as follows: bird 1: (0.279, 0.071), bird 2: (0.234, –0.885), bird 3: (0.177, 0.410), bird 4: (–0.495, 0.386), bird 5: (–1.219, 0.443). The updated positions of the birds are shown in Fig. 3.3.

Step 4: *Repeat step 2 and 3 till convergence*

Fig. 3.3 shows that, after first iteration, the swarm is at better position than earlier as it is evident from better average fitness value of 10.13 for updated positions than that of 11.35 for initial positions. Also it can be seen that in the updated swarm, bird 5 is at global best position (*gBest*) with fitness value of 5.24 whereas for initial population, bird 1 was at global best position (*gBest*) with fitness value of 8.6. Thus the global best value is also improved. For every bird in the swarm, the updated position is compared with its *pBest* value and will be updated as shown in Table 3.1.

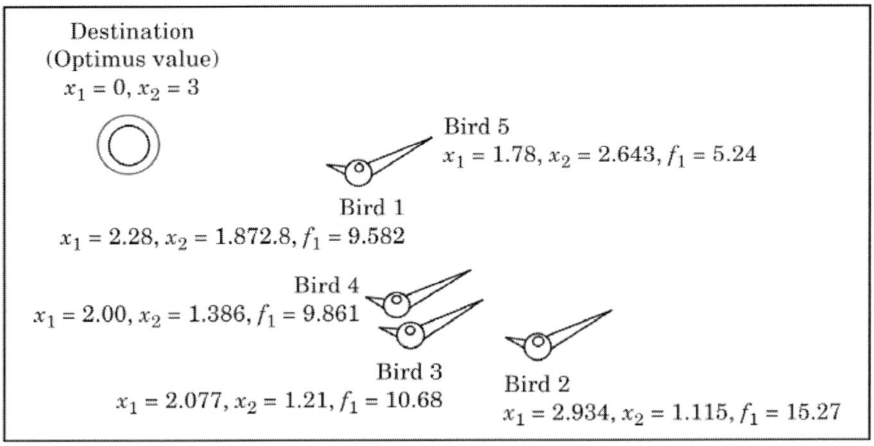

Fig. 3.3: Updated position of birds (first iteration).

Table 3.1: Updating *pBest* values.

Bird no.	***Updated positions and fitness values***	***Existing pBest values and fitness values***	***Updated pBest values***
1	$(x_1 = 2.28, x_2 = 1.87), f_1 = 9.58$	$(x_1 = 2, x_2 = 1.8), f_1 = 8.6$	$x_1 = 2, x_2 = 1.8$
2	$(x_1 = 2.93, x_2 = 1.11), f_2 = 15.27$	$(x_1 = 2.7, x_2 = 2), f_2 = 11.29$	$x_1 = 2.7, x_2 = 2$
3	$(x_1 = 2.08, x_2 = 1.21), f_3 = 10.68$	$(x_1 = 1.9, x_2 = 0.8), f_3 = 11.21$	$x_1 = 2.08, x_2 = 1.21$
4	$(x_1 = 2, x_2 = 1.39), f_4 = 9.86$	$(x_1 = 2.5, x_2 = 1), f_4 = 13.25$	$x_1 = 2, x_2 = 1.39$
5	$(x_1 = 1.78, x_2 = 2.64), f_5 = 5.24$	$(x_1 = 3, x_2 = 2.2), f_5 = 12.4$	$x_1 = 1.78, x_2 = 2.64$

As shown in Table 3.1, for bird 1 since current *pBest* is better (f_1 = 8.6) than its updated position (f_1 = 9.58), *pBest* will remain same. Similar is the case for bird 2. However, for birds 3, 4 and 5, as their updated positions are better than current *pBest* values, new *pBest* values will be their updated positions. Now the steps 2 and 3 will be repeated till convergence. At convergence, all birds will reach to the desired destination (optimum value) as shown in Fig. 3.4.

Applications of particle swarm optimization algorithm to deal with various issues related to manufacturing are presented through case studies 3.3, 3.4 and 3.5.

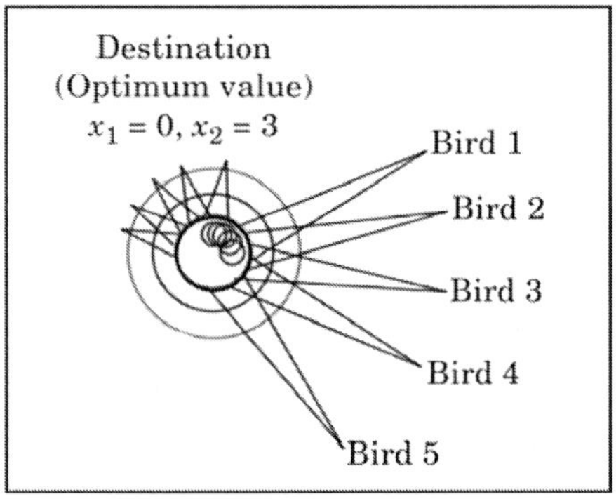

Fig. 3.4: Convergence at optimum solution.

3.3. Optimization of Abrasive Water Jet Machining (AWJM) Process Using PSO

The abrasive water jet machining (AWJM) process uses a high velocity water jet in combination with abrasive particles for cutting different types of materials using a setup as shown in Fig. 3.5.

A stream of small abrasive particles is introduced and entrained in the water jet in such a manner that water jet's momentum is partly transferred to the abrasive particles. The process thus combines the benefit of the two other advanced machining processes namely, water jet machining (WJM) and abrasive jet machining (AJM). This process relies on erosive action of an abrasive laden water jet for applications of cutting, drilling, and general cleaning and descaling of thick sections of very soft to very hard materials at high rates. Visual examination of the cutting process in AWJM suggests two dominant modes of material removal. First is erosion by cutting wear due to particle impact at shallow angles on the top surface of the kerf. Second is deformation wear due to excessive plastic deformation caused by particle impact at large angles, deeper into the kerf (Hashish, 1989). A variety of materials can be machined by AWJM including copper and its alloys, aluminum, lead, steel, tungsten carbide, titanium, ceramics, composites, acrylic, concrete, rocks, graphite, silica glass, etc. Its most promising application includes machining of sandwiched honeycomb structural materials frequently used in the aerospace industries. The main advantages of the process include omni-directional cutting, minimal thermal damage, no burrs, high cutting speed and efficiency, suitability for automation, and cutting without delaminating.

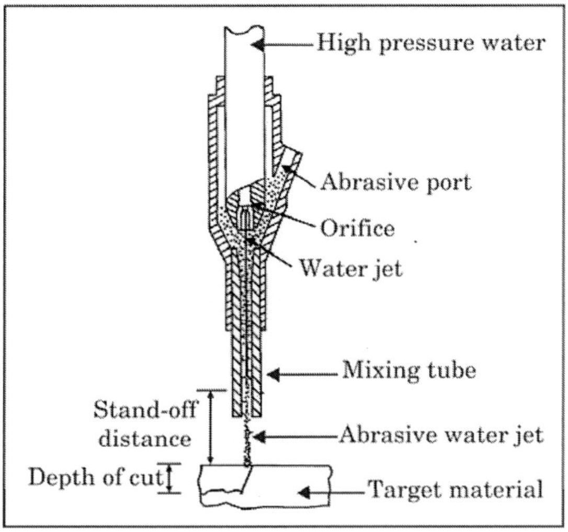

Fig. 3.5: AWJM process (Jain and Jain, 2001).

The optimization model for abrasive water jet machining process is formulated in the present work based on the analysis given by Hashish (1989). The five decision variables considered for this model are, water jet pressure at the nozzle exit (P_w), diameter of abrasive water jet nozzle (d_{awn}), feed rate of nozzle (f_n), mass flow rate of water (M_w), and mass flow rate of abrasives (M_a). The objective function and constraint for this process are discussed below:

3.3.1. *Objective function*

The objective is to maximize the material removal rate (Z_1) as given by equation (3.3).

$$Maximize\ Z_1 = d_{awn} f_n (h_c + h_d) \tag{3.3}$$

Where, 'h_c' is the indentation depth due to cutting wear as given by equation (3.4).

$$h_c = \left(\frac{1.028 \times 10^{4.5} \xi}{C_k^{1/3} f_r^{0.4}}\right)\left(\frac{d_{awn}^{0.2} M_a^{0.4}}{f_n^{0.4}}\right)\left(\frac{M_w P_w^{0.5}}{M_a + M_w}\right) - \left(\frac{18.48 K_a^{2/3} \xi^{1/3}}{C_k^{1/3} f_r^{0.4}}\right)\left(\frac{M_w P_w^{0.5}}{M_a + M_w}\right)^{1/3} \text{ if } \alpha_t \leq \alpha_0. \tag{3.4}$$

$$h_c = 0, \text{ if } \alpha_t \geq \alpha_0. \tag{3.5}$$

'h_d' is the indentation depth due to deformation wear as given by Eq. (3.6).

$$h_d = \frac{\eta_a d_{awn} M_a \left[K_1 M_w P_w^{0.5} - (M_a + M_w) v_{ac}\right]^2}{(1570.8\sigma_{fw}) d_{awn}^2 f_n (M_a + M_w)^2 + (K_1 C_{fw} \eta_a) \left[K_1 M_w P_w^{0.5} - (M_a + M_w) v_{ac}\right] M_a M_w P_w^{0.5}} \tag{3.6}$$

$$\alpha_0 \approx \left(\frac{0.02164 C_K^{1/3} f_r^{0.4}}{K_a^{2/3} \xi^{1/3}}\right)\left(\frac{\dot{M}_a + \dot{M}_w}{\dot{M}_w P_w^{0.5}}\right)^{1/3} \text{ (degrees)}. \tag{3.7}$$

$$\alpha_t \approx \left(\frac{0.389 \times 10^{-4.5} \rho_a^{0.4} C_K}{\xi}\right)\left(\frac{d_{awn}^{0.8} f_n^{0.4}\left(\dot{M}_a + \dot{M}_w\right)}{M_a^{0.4} \dot{M}_w P_w^{0.5}}\right) \text{ (degrees)}. \tag{3.8}$$

$$v_{ac} = 5\pi^2 \frac{\sigma_{cw}^{2.5}}{\rho_a^{0.5}}\left[\frac{1 - v_a^2}{E_{Ya}} + \frac{1 - v_w^2}{E_{Yw}}\right]^2 \text{ (mm/s)} \tag{3.9}$$

$$K_1 = \sqrt{2} \times 10^{4.5} \xi. \tag{3.10}$$

$$C_K = \sqrt{3000\, \sigma_{fw} f_r^{0.6} / \rho_a} \text{ (mm/s)} \tag{3.11}$$

$$K_a = 3.$$

3.3.2. *Constraint*

Constraint is on power consumption as given by Eq. (3.12).

$$1.0 - \frac{P_w M_w}{P_{max}} \geq 0 \tag{3.12}$$

Description of various symbols appeared in equations (3.4) to (3.12) is provided below.

Notation	***Description***	***Unit***	***Value***
ρ_a	Density of abrasive particles	Kg/mm^3	3.95×10^{-6}
υ_a	Poisson ratio of abrasive particles		0.25
E_{Ya}	Modulus of elasticity of abrasive particles	MPa	350000
f_r	Roundness factor of abrasive particles		0.35
f_s	Sphericity factor of abrasive particles		0.78
η_a	Proportion of abrasive grains effectively participating in machining		0.70
υ_w	*Poisson* ratio of work material		0.20
E_{Yw}	Modulus of elasticity of work material	MPa	114000
σ_{ew}	Elastic limit of work material	MPa	883
σ_{ew}	Flow stress of work material	MPa	8142
C_{fw}	Drag friction coefficient of work material		0.002
ξ	Mixing efficiency between abrasive and water		0.8
P_{max}	Allowable power consumption value	kW	56

3.3.3. *Variable bounds*

The bounds for the five variables are given in equations (3.13) to (3.17).

$$50 \leq P_w \leq 400 \text{ (MPa)} \tag{3.13}$$

$$0.5 \leq d_{awn} \leq 5 \text{ (mm)} \tag{3.14}$$

$$0.2 \leq f_n \leq 25 \text{ (mm/s)} \tag{3.15}$$

$$0.02 \leq M_w \leq 0.2 \text{ (Kg/s)} \tag{3.16}$$

$$0.0003 \leq M_a \leq 0.08 \text{ (Kg/s)} \tag{3.17}$$

Now, the steps of PSO algorithm discussed in section 3.2 are implemented for optimization of AWJM process. The optimum selection of operating parameters of PSO algorithm like acceleration constants 'c_1' and 'c_2' as well as inertia coefficient 'w' is very essential for convergence of the algorithm. Considering the velocity and positions of a particle at discrete time steps, the following non-homogeneous recurrence relation is obtained (Bergh and Engelbrecht, 2006).

$$X_{i+1} = (1 + w - \phi_1 - \phi_2)X_i - wX_{i-1} + \phi_1 \times pBest_i + \phi_2 \times gBest_i \tag{3.18}$$

Where, $\phi_1 = c_1 \times r_1$ and $\phi_2 = c_2 \times r_2$

This recurrence relation can be written in matrix-vector notation as the product:

$$\begin{bmatrix} X_{i+1} \\ X_i \\ 1 \end{bmatrix} = \begin{bmatrix} 1+w-\phi_1-\phi_2 & -w & \phi_1 \times pBest_i + \phi_2 \times gBest_i \\ 1 & 0 & 0 \\ 0 & 0 & 1 \end{bmatrix} \begin{bmatrix} X_i \\ X_{i-1} \\ 1 \end{bmatrix} \tag{3.19}$$

The characteristics polynomial of the matrix in equation (3.19) is

$$(1 - \lambda)\,(\mathrm{w} - \lambda\,(1 + w - \phi_1 - \phi_2) + \lambda^2) \tag{3.20}$$

The solution to this polynomial gives Eigen values.

$$\lambda_1 = \frac{1+w-\phi_1-\phi_2+\gamma}{2} \tag{3.21}$$

$$\lambda_2 = \frac{1+w-\phi_1-\phi_2-\gamma}{2} \tag{3.22}$$

Where, $$\gamma = \sqrt{(1+w-\phi_1-\phi_2)^2 - 4w} \tag{3.23}$$

Now, to ensure the convergence of algorithm, the values of λ_1 and λ_2 should be such that *max*

$$(|l_1|,\ |l_2|) < 1 \tag{3.24}$$

This can be achieved only when the condition given by equation (3.25) is satisfied

$$w > 0.5\ (\phi_1 + \phi_2) - 1 \tag{3.25}$$

Considering several trial combinations of w, c_1, and c_2, it is observed that the condition given by Eq. (3.25) is violated for all combinations of w, c_1, and c_2 falling in the shaded region shown in Fig. 3.6. Hence this region is a strictly divergence zone.

As the feasible range for w is 0–1 and for c_1 and c_2 is 0–2, the selected values of w, c_1 and c_2 should be such that the equation (3.25) is satisfied for all possible values of random numbers r_1 and r_2 in the range 0–1. Keeping in view of this, considerable number of trials are conducted and the values of w, c_1 and c_2 are finally selected as 0.65, 1.65 and 1.75 respectively. With this parameter setting, the optimum process parameter values obtained by using PSO algorithm are given in Table 3.2.

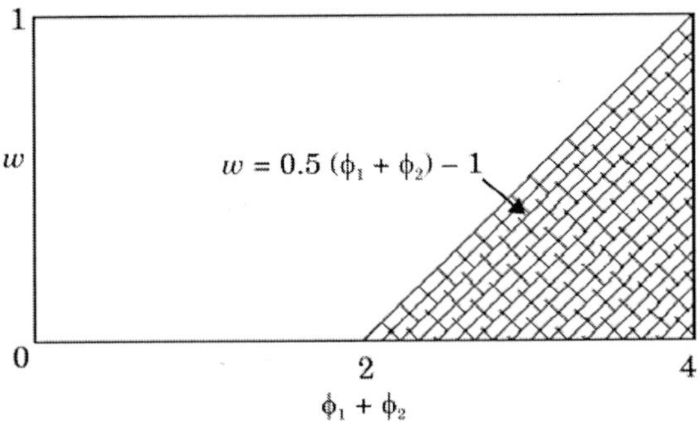

Fig. 3.6: Convergence criteria of PSO algorithm (Bergh and Engelbrecht, 2006).

Table 3.2: Optimum process parameter values obtained using PSO algorithm.

d_{awn} (mm)	f_n (mm/s)	M_w (Kg/s)	P_w (MPa)	M_a (Kg/s)	a_0 (°)	a_t (°)	h_c (mm)	h_d (mm)	MRR (mm³/s)	Power (kW)
3.242	13.084	0.14	400	0.08	0.385	0.384	3.31	2.12	230.50	56

Table 3.2 shows the comparative performance of particle swarm optimization algorithm PSO along with that reported in literature by using genetic algorithm (Jain *et al.,* 2007).

Table 3.3: Results of optimization using GA and PSO.

Method	d_{awn} (mm)	f_n (mm/s)	M_w (Kg/s)	P_w (MPa)	M_a (Kg/s)	a_0 (°)	a_t (°)	h_c (mm)	h_d (mm)	MRR (mm³/s)	Power (kW)
GA (Jain *et al.*, 2007)	3.726	23.17	0.141	398.3	0.079	0.384	0.572	0.00	1.045	90.257	56
PSO	3.242	13.084	0.140	400	0.080	0.385	0.384	3.31	2.12	230.50	56

For abrasive water jet machining, if angle of impingement at the top of the machined surface 'a_t' exceeds the critical impact angle 'a_0' then no material removal is assumed to occur by cutting wear ($h_c = 0$) and the material removal takes place only due to the deformation wear, resulting into relatively less material removal rate (Paul *et al.,* 1998). As shown in Table 3.3, for the solution obtained by using genetic algorithm, as 'a_t' exceeds 'a_0', indentation depth of cutting wear (h_c) becomes zero which results in less material removal rate as compared to the solution obtained by using particle swarm optimization algorithm for which 'a_t' < 'a_0'. The optimum values of process variables obtained by using particle swarm optimization algorithm also offers higher value of depth of deformation wear (h_d) than that obtained by using genetic algorithm, which further increases the material removal rate. The combined effect thus leads to the improvement in objective function by about 155 %.

However, if the parameter are set such that they don't satisfy the convergence criteria *viz.* w = 0.2, c_1 = 1.90 and c_2 = 1.95, then it is observed from Table 3.4, that when all other conditions such as population size, initial solutions, and number of generations etc. are kept same, the results of PSO algorithm are inferior to those obtained using genetic algorithm.

Table 3.4: Results of optimization using GA and PSO.

Method	d_{awn} *(mm)*	f_n *(mm/s)*	M_w *(Kg/s)*	P_w *(MPa)*	M_a *(Kg/s)*	*MRR* *(mm³/s)*	*Power* *(kW)*
GA (Jain *et al.*, 2007)	3.726	23.17	0.141	398.3	0.079	90.25	56
PSO	2.41	4.15	0.20	267.1	0.025	72.08	56

This examples thus demonstrates the fact that the performance of the PSO algorithm is governed by selection of its algorithm specific parameters w, c_1, and c_2.

3.4. Optimization of Control Parameters of Cooling System for an Industrial Robot Controller

It is a well-known fact that all equipments applying electronic parts (such as semiconductors, control cards, VFD, etc.) are severally affected by heat generated which leads to malfunction and shortened life of the device. Hence, automation control systems allowing electronic components to run hot can be a costly gamble. Provision of cooling system is the most effective way of protecting the electronic components from overheating and thereby increasing their life. However, cooling systems are faced with many problems in actual practice and undesirable results affect their efficiency and performance. Problems such as excessive condensation, over–flow of drain tray and excessive power consumption are the most common problems faced in the cooling systems used for automation control systems in industries. An industrial case study is discussed in this section to demonstrate the application of PSO algorithm to improve the performance of a cooling system (an air conditioning unit) in terms of power consumption and cooling efficiency for robot controller.

The control parameters considered for the air conditioning unit are temperature, air flow, and air pressure. Their operating ranges are shown in Table 3.5.

Table 3.5: Operating range of parameters selected for study.

Parameter	*Notation*	*Range*	*Coded values*				
			–2	*–1*	*0*	*1*	*2*
Temperature	x_1	20 to 32 (°C)	20	23	26	29	32
Air flow	x_2	5.5 to 11.5 (m/sec)	5.5	7	8.5	10	11.5
Air pressure	x_3	8 to 20 (bar)	8	11	14	17	20

The objective is to minimize the electric power consumption in kWh (z_1) while ensuring that necessary heart transfer rate (z_2) will be maintained. The model for z_1 is a second order RSM model as given by Eq. (3.26) (Nighot and Pawar, 2017):

$$z_1(kWh) = 2.4551 - 0.3842\,x_1 - 0.0929\,x_2 - 0.0769\,x_3 + 0.0306\,x_1^2 + 0.1480\,x_2^2 - 0.0187\,x_3^2 + 0.0585\,x_1.x_2 - 0.0217\,x_1.x_3 + 0.0044\,x_2.x_3 \quad (3.26)$$

The constraint function z_2 can be formulated using basic heat transfer equation:

$$Q\ (J/sec) = \frac{m \times C \times \Delta T}{t} \quad (3.27)$$

Where, Q = Total amount of heat to be removed by cooling substance = 984 J/s (this is calculated from total amount of heat generated by all electronics parts of the controller such as variable frequency drive (VFD), transformer etc.)

m = Mass of the cooling substance (R134a) = 0.5 Kg

C = Specific heat capacity of cooling substance = 1005 J/kg K

ΔT = Initial temperature (before cooling) – Final temperature (after cooling) = 50 - x_1

t = Cooling time (Sec).

$$\frac{Volume\ of\ enclosure\ (m^3)}{Air\ flow\ rate\ \left(\frac{m^3}{s}\right) \times crosss\ section\ of\ blower\ duct\ (m^2)} = \frac{0.4243}{x_2 \times 0.004} = \frac{106}{x_2}$$

$$\text{Thus, } Q = \frac{0.5 \times 1005 \times (50 - x_1) \times x_2}{106} = 4.74 \times (50 - x_1) \times x_2$$

Now, to ensure that all heat generated to be transferred by cooling system, the constraint equation is formulated as given by Eq. (3.28).

$$4.74 \times (50 - x_1) \times x_2 \geq 984$$

$$z_2 = 4.74 \times (50 - x_1) \times x_2 \geq 984$$

The combined objective function is thus formulated as:

$$Minimize\ z = z_1 - penalty \times z_2 \tag{3.28}$$

Penalty is calculated using static penalty method = 0.1 if $z_2 < 0$ else, penalty = 0. It is to be noted that z_1 is calculated using coded values whereas z_2 is calculated using actual values. Penalty is so determined that it will take care of conversion from coded values to actual values.

The PSO algorithm is now applied to solve above model as discussed below:

Step 1: *Generate randomly initial population of solution*

Considering population size (*i.e.,* birds in swarm) to be 10, the initial positions of birds and their fitness values are given in Table 3.6. Initial *pBest* values are same as that of bird positions and 6th solution is *gBest*. Considering the PSO parameters as w = 0.65, c_1 = 1.65, c_2 = 1.75, and initial velocities are shown in Table 3.6.

Table 3.6: Initial population of solution.

Solution no.	x_1	x_2	x_3	z	*Initial pBest* (x_1, x_2, x_3)	*Initial velocity* (x_1, x_2, x_3)
1	1.139	1.429	–1.149	2.407	(1.139,1.429,–.149)	(1.407, 0.830, 0.692)
2	1.515	2	–1.737	2.646	(1.515, 2.519, –1.737)	(1.247, 1.201, 0.111)
3	1.792	1.243	0.826	8.909	(1.792, 1.243, 0.826)	(0.286, 0.013, 1.185)
4	0.772	2	–0.959	2.737	(0.772, 2.423, –0.959)	(0.259, 1.241, 0.703)
5	1.902	1.717	1.224	4.500	(1.902, 1.717, 1.224)	(1.305, 1.374, 0.710)
6 (*gBest*)	0.852	1.04	–0.564	2.311	(0.852, 1.040, –0.564)	(0.390, 0.773, 1.311)
7	1.541	1.539	–1.067	2.371	(1.541, 1.539, –1.067)	(1.089, 0.783, 0.156)
8	0.058	0.392	0.264	2.399	(0.058, 0.392, 0.264)	(0.290, 0.575, 1.293)
9	1.524	0.73	–1.093	12.152	(1.524, 0.730, –1.093)	(1.017, 0.002, 0.601)
10	1.723	1.853	–0.801	2.480	(1.723, 1.853, –0.801)	(1.086, 0.980, 1.466)

Step 2: *Identify pBest and gBest*

Initial *pBest* value for each bird is its current position. As the problem is of minimization, 6th solution is having best fitness value, hence is *gBest*.

Step 3: *Update positions of birds*

For the initial velocities shown in Table 3.6 and considering the three coefficients as: Inertia coefficient (w) = 0.65, cognitive parameter (c_1) = 1.65, and social parameter (c_2) = 1.75, Eq. (3.1) and Eq. (3.2) are used to determine new velocities and new positions of all birds. The updated velocities and positions of the birds are shown in Table 3.7.

Step 4: *Repeat steps 2 and 3 till convergence*

It can be seen from Table 3.7 that after first iteration the swarm is having better position than earlier as it is evident from better average fitness value of 3.373 for updated position than that of 4.291 for initial position. Also it can be seen that the in updated swarm, bird 4 is at global best position (*gBest*) with fitness value of 2.150 whereas for initial population, bird 6 was at global best position (*gBest*) with fitness value of 2.311. Thus the global best value is also improved. For every bird in the swarm, the updated position is compared with its *pBest* value which will be updated as shown in Table 3.7. The step 2 and 3 will be repeated till convergence criterion is met.

Table 3.7: Updated velocities and positions of the birds.

Solution no.	x_1	x_2	x_3	z	*Updated pBest* (x_1, x_2, x_3)	*Updated velocity* (x_1, x_2, x_3)
1	1.112	1.951	0.312	2.544	(–1.868, 1.470, –0.062)	(–0.027, 0.522, 1.461)
2	–1.155	0.071	0.600	2.891	(–0.666, –0.241, 0.669)	(–2.670, –1.929, 2.337)
3	–2.000	1.535	1.188	3.314	(–2.000, 1.535, 1.188)	(–4.878, 0.292, 0.362)
4 (*gBest*)	1.302	1.727	2.000	2.150	(1.302, 1.727, 2.000)	(0.530, –0.273, 4.448)
5	2.000	2.000	–1.008	2.825	(2.000, 2.000, –1.008)	(0.761, 0.489, –2.232)
6	1.151	1.872	2.000	2.262	(1.051, 1.382, 0.707)	(0.299, 0.832, 3.426)
7	1.122	2.000	–0.457	2.638	(0.939, –0.004, –1.983)	(–0.419, –0.296, 0.610)
8	1.720	1.190	0.919	8.536	(0.878, 0.034, –0.686)	(1.662, 0.798, 0.655)
9	2.000	1.860	2.000	4.138	(2.000, 1.860, 2.000)	(–3.953, 1.130, 3.757)
10	2.000	2.000	2.000	2.433	(–1.749, 0.206, 1.304)	(–2.554, -2.026, 5.759)

The convergence graph for 25 iterations is shown in Fig. 3.7.

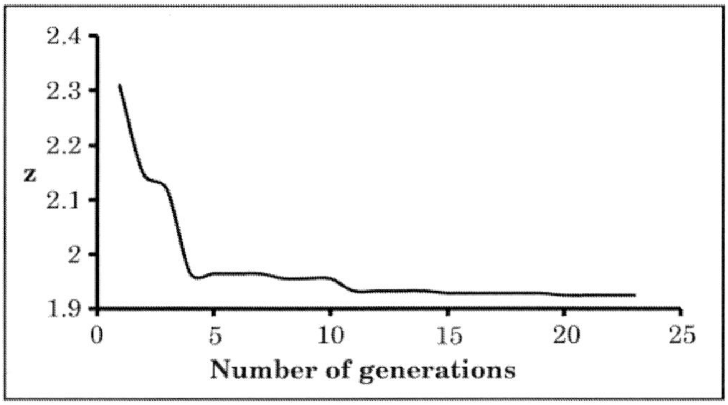

Fig. 3.7: Convergence graph

The results of optimization using PSO algorithm for the present case study are shown in Table 3.8.

Table 3.8: Result of optimization for control parameters of cooling System using PSO algorithm.

Coded values			*Actual values*			*Electric power consumption*	*Heat transfer rate*	*Combined objective function*
x_1	x_2	x_3	Temperature	Air flow	Air pressure	z_1	z_2	z
1.168	1.087	2.000	29.504°C	10.131 (m/sec)	20 bar	1.926 kwh	984.19 J/s	1.926

Application of PSO algorithm to optimize the performance of cooling system of robot controller thus proved to be an effective approach resulting into significant improvement of about 50% over manual recommendations.

3.5. Modelling and Optimization of Process Parameters of Electric Discharge Machining to Minimize Wire Breakage

In recent years an increasing demand for machining of complex shapes made of hard and difficult-to-machine materials with exact tolerances and surface finish resulted in the development of many advanced machining processes based on chemical, electro-chemical, thermal, electro-thermal, mechanical and other means of material removal. Wire electric discharge machining (WEDM) is one of the widely accepted advanced machining processes used to machine the components with intricate shapes and profiles. It is considered as a unique adaptation of the conventional EDM process which uses an electrode to initialize the sparking process. WEDM utilizes a continuously travelling wire electrode made of thin copper, brass or tungsten. On application of a proper voltage, discharge occurs between the wire electrode and the work-piece in the presence of a flood of deionized water of high insulation resistance. The material is eroded ahead of the wire through a series of repetitive sparks between electrodes, *i.e.,* work-piece and the wire.

WEDM has been gaining wide acceptance in modern tooling applications, machining of advanced ceramic materials and modern composite materials due to following reasons:

- As the wire diameter is small (0.05–0.3 mm), the process is capable of achieving very small corner radii.
- The wire is kept in tension using a mechanical tensioning device reducing the tendency of producing inaccurate parts.
- During the WEDM process there is no direct contact between the work-piece and the wire, eliminating the mechanical stresses during machining.
- WEDM process is able to machine exotic and high strength and temperature resistive (HSTR) materials and eliminate the geometrical changes occurring in the machining of heat-treated steels.

Wire electric discharge machining (WEDM) is thus an important non-traditional machining process, specially used to make complex and intricate shaped components from difficult to machine materials. However, wire breakage is one of the major issues frequently encountered while wire electric discharge machining of hard materials like EN19 commonly used for defense applications. While machining, if wire breaks, it will stop the operation in serious situation. After wire breakage all the operation settings has to be recovered and broken wire has to be replaced. It is difficult to restart the machining operation from the same work-piece and tool alignment settings. Thus, this replacement increases the machining time. Wire breakage also produces a permanent surface damaged on the component surface and also decreases accuracy of the process. This will greatly hamper the machining performance and productivity of WEDM. There are various causes of wire breakage in WEDM including the thermal loading, high wire tension due to intricate and complex shape of components, impact of electric discharge and large vibrations of wire, etc. The machining parameters that affects material removal rate and wire breakage includes pulse on time, pulse off time, peak voltage, wire feed, flushing pressure, cutting speed, etc. For the high electric discharge higher pulse on time is required to select which causes the short circuit and wire breakage. As the pulse on time increases the electric discharge rate increases that will lead to the wire breakage. The peak voltage directly affects the wire breakage. Due to the higher value of peak voltage heat energy increases that leads to melting and vaporization of wire. The selection of proper wire feed rate results into higher material removal rate, less wire breakage and better surface finish. Flushing pressure refers to the pressure of flushing fluid used to remove the molten slug. The high flushing pressure causes the wire deflection and vibrations, but it is also necessary for better cutting performance. Material removal rate (MRR) is also an important aspect in WEDM to achieve higher machining productivity and cost effective utilization of machine. While selecting appropriate parameters for minimum frequency of wire breakage care has to be taken that material removal rate should not be hampered. As these two objectives *i.e.*, minimization of frequency of wire breakage and maximization of material removal rate are conflicting in nature, selection of appropriate process parameters is a challenging task.

An effective way to solve this problem is to discover the relationship between the performance of the process and its controllable input parameters by modeling the process through suitable mathematical techniques and optimization using suitable optimization algorithm. In the present work, response surface methodology is used to model the process whereas the optimum parameter setting is achieved by employing particle swarm optimization (PSO) algorithm.

Response surface modeling (RSM)

Response surface modeling (RSM) is a collection of statistical and mathematical methods that are useful for modeling and optimization of the engineering science problems. RSM quantifies the relationship between the controllable input parameters and the obtained responses. In modeling of manufacturing processes using RSM, sufficient data is collected through designed experimentation. An experiment is designed with 2^k (where, k = number of variables, in this study k = 7) factorial with central composite-second order ratable design is used. This consists of number of corner points =128, number of axial points=14, and a center point at zero level =7. The axial points are located in a coded test condition space through parameter '*a* '. For the design to remain rotatable, '*a* ' is taken as 2. Thus the coded level for the axial points is at 2. The center point is repeated four times to estimate the pure error. The coded value corresponding to actual value for each process variable is derived using the following formula:

$$\text{Coded test condition} = \frac{\text{Actual test condition - mean test condition}}{\text{Range of test condition/2}}$$

Table 3.9 shows coded values of process variables.

Table 3.9: Coded values of process variables.

Parameters	*Coded levels*				
	–2	*–1*	*0*	*1*	*2*
Pulse on time (T_{on}), Sec	103	105	107	109	111
Pulse off time (T_{off}), Sec	23	30	37	44	51
Peak voltage (V_p), Volts	10	11	12	13	14
Water pressure (P_w) Kgf/cm^2	7	11	15	19	23
Wire feed (f), mm/min	2	3	4	5	6
Machine speed (V_m), RPM	75	100	125	150	175
Cutting speed (V_c), %	15	20	25	30	35

The details of experimental set up used for data collection are given below:

- Machine type = ULTIMA 1S/2S (Wire cut WEDM submerged type)
- Pulse generator = ELPULS 55S SERIES
- Make = Electronica Machine tools
- Wire material = Brass diffusion coated
- Wire diameter = 0.25 mm
- Wire composition = Cu 65% and Zn 35%
- Wire strength = 400 N/mm^2
- Wire tension = 19.63 Newton
- Dielectric = distilled water (HCL and Carbide soda)
- Range of dielectric temperature = 20°C to 240°C
- Work piece material = 40Cr4Mo3
- Work piece specification = special tool for factory use

To study the effect of process parameters *i.e.,* T_{on}, T_{off}, V_p, P_w, f, V_m, and V_c on performance measures *i.e.,* Material removal rate (*MRR*) and wire breakage in terms of tension of used wire (*TENSION*), a second-order polynomial response is fitted into the following equation.

$$y = b_0 + \sum_{i=1}^{k} b_i x_i + \sum_{i=1}^{k} b_{ii} x_i^2 + \sum_{j>1}^{k} b_{ij} x_i x_j \tag{3.29}$$

Where 'y' is the response and the x_i (1, 2... k) are coded levels of k quantitative variables. The coefficient b_0 is the free term, the coefficients b_i are the linear terms, the coefficients b_{ii} are the quadratic terms, and the coefficients b_{ij} are the interaction terms. Equations (3.30) and (3.31) are then derived by determining the values of the coefficients using the least square technique for the observations collected.

MRR = 0.8486 + 0.0477 × x_1 + 0.058 × x_2 + 0.066 × x_3 + 0.043 × x_4 +0.033 × x_5 + 0.0088 × x_6 + 0.0044 × x_7 – 0.0026 × x_1^2 – 0.0119 × x_2^2 – 0.0244 × x_3^2 – 0.0026 × x_4^2 0.015 × x_5^2 + 0.0005 × x_6^2+0.050 × x_7^2 – 0.0015 × x_1 × x_2 – 0.0015 × x_1 × x_3 + 0.0015 × x_1 × x_4 – 0.00078 × x_1 × x_5 + 0.00078 × x_1 × x_6 + 0.00156 × x_1 × x_7 – 0.00156 × x_2 × x_3 + 0.00078 × x_2 × x_4 – 0.0015 × x_2 × x_5 + 0 × x_2 × x_6 + 0.00156 × x_2 × x_7 + 0.00156 × x_3 × x_4 – 0.00078 × x_3 × x_5 – 0.00078 × x_3 × x_6 + 0.00078 × x_3 × x_7 +

$0 \times x_4 \times x_5 + 0 \times x_4 \times x_6 - 0.0015 \times x_4 \times x_7 - 0.0054 \times x_5 \times x_6 - 0.0039 \times x_5 \times x_7 - 0.0023 \times x_6 \times x_7$ (3.30)

$$TENSION = 3.079 - 0.0845 \times x_1 - 0.1514 \times x_2 - 0.2536 \times x_3 - 0.086 \times x_4 + 0.0463 \times x_5 - 0.04485 \times x_6 - 0.0720 \times x_7 - 0.0099 \times x_1^2 + 0.0463 \times x_2^2 + 0.0525 \times x_3^2 - 0.0099 \times x_4^2 - 0.01617 \times x_5^2 - 0.0099 \times x_6^2 + 0.0244 \times x_7^2 + 0.0093 \times x_1 \times x_2 - 0.00313 \times x_1 \times x_3 + 0.0039 \times x_1 \times x_4 - 0.0101 \times x_1 \times x_5 + 0.0046 \times x_1 \times x_6 + 0.0031 \times x_1 \times x_7 - 0.0031 \times x_2 \times x_3 - 0.0085 \times x_2 \times x_4 - 0.0031 \times x_2 \times x_5 - 0.0054 \times x_2 \times x_6 + 0 \times x_2 \times x_7 + 0.007 \times x_3 \times x_4 + 0.0007 \times x_3 \times x_5 + 0.0046 \times x_3 \times x_6 + 0.0085 \times x_3 \times x_7 + 0.0031 \times x_4 \times x_5 + 0.0093 \times x_4 \times x_6 + 0.0085 \times x_4 \times x_7 + 0.0007 \times x_5 \times x_6 + 0.0062 \times x_5 \times x_7 + 0.0117 \times x_6 \times x_7 \quad (3.31)$$

The steps of PSO algorithm as discussed in section 3.2 are implemented to solve this optimization model and the results of optimization are as shown in Table 3.10.

Table 3.10: Results of optimization of process parameters of wire electric discharge machining to minimize wire breakage.

Parameters	***Optimum values***
Pulse on time (T_{ON}), μSec	103.63
Pulse off time (T_{OFF}), μSec	35.11
Peak voltage (V_P), Volts	10.5
Water pressure (P_W), Kgf/cm^2	8.508
Wire feed (f), mm/min	5.25
Machine speed (V_M), RPM	103.17
Cutting speed override (O), %	29.7

By setting above mentioned optimum parameters on WEDM machine, material removal rates and wire breakage are measured for 120 runs and it is observed that the average MRR is 1.0506 with no wire breakage. It is also observed that for these 120 readings, the standard deviation for MRR values is less than 10%, which indicate the robustness of the solution.

3.6. Variants of PSO Algorithm

Although, particle swarm optimization (PSO) algorithm has been successfully applied to many real life problems, it suffers from a drawback of getting stuck in the local minima. To improve the performance of PSO algorithm, the researchers had proposed several variants of PSO algorithm.

3.6.1. *Opposition based learning PSO*

As we have seen that PSO starts with some initial solutions (initial swarm) and tries to improve performance toward some optimal solutions. The process of searching terminates when predefined criteria are satisfied. In absence of a priori information about the solution, we start with a random guess. Obviously, the computation time is directly related to distance of the guess from optimal solution. We can improve our chance to start with a closer (fitter) solution by checking the opposite solution simultaneously. By doing so, the closer one to solution (say guess or opposite guess) can be chosen as initial solution. In fact, according to probability theory, in 50% of cases the guess is farther to solution than opposite guess; for these cases starting with opposite guess can accelerate convergence. The same approach can be applied not only to initial solutions but also continuously to each solution in the current swarm. The concept of opposition-based learning was introduced by Tizhoosh (2005). The opposite number $\widetilde{x_i}$ of any variable x_i can be defined as $\widetilde{x_i} = L_i + U_i - x_i$, where L_i and U_i are lower and upper bounds of variable i respectively. Thus, two sets of initial solutions are obtained, one with original guess x_i and other with opposite guess . The best out of x_i and $\widetilde{x_i}$ is considered as final initial guess. Also in PSO, the initial velocity (v_i) is also a guess, the opposite velocity $\widetilde{v_i}$ is also computed using Eq. (3.32).

$$\widetilde{v_i} = V_{i\_max} + V_{i\_min} - v_i \tag{3.32}$$

Where, V_{i_max} and V_{i_min} are maximum and minimum velocities of the population respectively.

Table 3.11: Opposite of initial solutions and velocities for objective function mentioned in section 3.2.

x_1	x_2	f_i	v_1	v_1	$\widetilde{x_1}$	$\widetilde{x_2}$	$\widetilde{f_i}$	$\widetilde{v_1}$	$\widetilde{v_2}$
2	1.8	8.6	0.43	0.11	1	1.2	7.4	0.94	1.87
2.7	2	11.29	1.36	0.49	0.3	1	7.09	0.01	1.49
1.9	0.8	11.21	0.01	0.42	1.1	2.2	4.61	1.36	1.56
2.5	1	13.25	0.33	1.87	0.5	2	4.25	1.04	0.11
3	2.2	12.4	0.07	1.03	0	0.8	7.6	1.3	0.95

In Table 3.11, and . It can be seen from Table 3.11, that all opposite solutions are better than original solutions. As we are starting with better set of initial solution, convergence can be accelerated.

Zhang *et al.* (2009) proposed an enhance variant of PSO which they called quasi-oppositional comprehensive learning particle swarm optimizers (QCLPSO). They used the qausi opposite numbers for the swarm initialization.

3.6.2. *Discrete PSO algorithm*

This algorithm is applied to discrete and combinatorial optimization problems over discrete-valued search space (Parsopoulos and Vrahatis, 2002; Kumar *et al.,* 2016). In this algorithm, particle's positions are discrete values. The velocity and position equation are developed for real values and updated in each iteration. Discrete PSO has a high success rate in solving integer programming problems as compare with other methods such as branch-and-bound. It has a quick convergence and better performance results.

3.6.3. *Civilized swarm optimization (CSO)*

Civilized swarm optimization (Selvakumar and Immanuel, 2009) is an integrating technique of PSO and Society civilization algorithm (SCA). In SCA proposed by Ray and Liew (2003), the solution is searched by the interaction of intra-society and inter-society. The mutually interacting individuals are grouped in societies (clusters) and the best-performing individual of a society is selected as society leader (SL). All societies construct the civilization and best SL is selected as civilization leader (CL). For optimization, individuals are following the SL *i.e.,* intra-society interaction and the SL is improved themselves by migrating to the CL of the civilization *i.e.,* intersociety. The perception behind inter and intra-society interaction are based on two fundamentals observations from nature.

1. A civilization emerges and advances due to cooperative relationships among its societies.
2. Individuals in a society interact with each other with an aim to improve.

The operating mechanism of SCA is differing with PSO, in SCA the individuals are following their society leader and neglecting their own experience. CSO hybridized the PSO and SCA in which the particle of the swarm is searched with the help of own experiences and the SL society to exploit a "promising area". Civilized swarm optimization is a powerful hybrid technique combining advantages of both SCA and PSO. Following are a few points of comparison:

- Society civilization algorithm (SCA) provides a well balance between local searching strategy and global search. It facilitates the exploration of promising area effectively and exploitation of these searched areas efficiently in a well-balanced manner while PSO has the capability of utilizing the personal experiences of the particles efficiently.
- In PSO, the gathered information regarding the search space is used successfully in deciding the optimal searching area whereas SCA fails to do so.
- Civilized swarm optimization (CSO) possesses a detailed knowledge of the searching space because of the integration of strengths of both SCA and PSO strategies.
- It provides a faster convergence rate and enhanced quality of results with proper satisfaction of constraints.
- Particle swarm optimization in comparison to SCA has the individuals obtaining the global best value considering their own experiences as well. This utilization of the personal experiences makes them to exploit a promising region by moving inside formed societies.

The application of CSO algorithm for application example discussed in section 3.4 is demonstrated through following steps:

Step 1: ***Determining algorithm specific parameters***

- Number of birds in swarm (Population size, N) = 20
- Number of societies ($N/_s$): 4
- Inertia weight (w) = 0.65
- Ognitive parameter (c_1): 1.65
- Social parameter (c_2): 1.75

Step 2: ***Generate initial population***

Initial 20 solutions along with their *pBest* values and velocities are shown in Table 3.12.

Table 3.12: Initial population of solutions for CSO algorithm.

Solution no.	x_1	x_2	x_3	z	*Initial velocity* (x_1, x_2, x_3)
1	1.139	1.429	–1.149	2.407	(1.407, 0.830, 0.692)
2	1.515	2	–1.737	2.646	(1.247, 1.201, 0.111)
3	1.792	1.243	0.826	8.909	(0.286, 0.013, 1.185)
4	0.772	2	–0.959	2.737	(0.259, 1.241, 0.703)
5	1.902	1.717	1.224	4.500	(1.305, 1.374, 0.710)
6	0.852	1.04	–0.564	2.311	(0.390, 0.773, 1.311)
7	1.541	1.539	–1.067	2.371	(1.089, 0.783, 0.156)
8	0.058	0.392	0.264	2.399	(0.290, 0.575, 1.293)
9	1.524	0.73	–1.093	12.152	(1.017, 0.002, 0.601)
10	1.723	1.853	–0.801	2.480	(1.086, 0.980, 1.466)
11	1.285	0.999	1.391	4.899	(0.615, 0.929, 1.098)
12	–1.391	1.065	1.640	2.912	(1.685, 1.595, 0.392)
13	0.854	0.672	1.443	3.809	(0.782, 0.629, 0.221)
14	–0.541	0.279	–1.771	2.703	(1.811, 0.714, 0.035)
15	–1.921	–1.351	0.711	10.888	(1.839, 0.695, 1.204)
16	0.211	1.415	0.673	2.498	(0.318, 1.138, 1.925)
17	1.401	1.465	–0.997	2.361	(1.619, 1.306, 0.981)
18	0.466	0.005	1.474	9.369	(0.818, 0.340, 1.791)
19	0.265	–1.667	–1.521	35.406	(1.984, 0.320, 0.953)
20	1.827	–1.990	0.234	52.376	(0.969, 0.218, 0.659)

Step 3: ***Arrange the individuals in ascending order according to their pBest values and decide CL and SL***

Since our problem is of minimization, the individuals are arranged in ascending order of fitness function of their *pBest* values as shown in Table 3.13. As discussed earlier, for initial population, *pBest* is the current solution.

The first solution (solution number 6) is civilization leader (CL). As number of societies are 4, the first four solutions *i.e.,* solution numbers 6, 17, 7, 8 are society leaders (SL).

Step 4: ***Assign the society members to society***

As there are four societies led by four society leaders, the members are assigned to particular society having minimum normalized Euclidean distance from corresponding society leader. The normalized Euclidean distance is computed by using Eq. (3.33).

$$d = \sqrt{\sum \frac{x_{sSL} - x_s}{x_s^{max} - x_s^{min}}} \tag{3.33}$$

where, x_s the value of sth decision variable, x_{sSL} is corresponding value of variable of society leader. and are upper and lower limits of the sth decision variable respectively.

Table 3.13: Individuals are arranged in ascending order of fitness values.

Solution no.	x_1	x_2	x_3	z	*Initial velocity* (x_1, x_2, x_3)
6	0.852	1.04	–0.564	2.311	(0.390, 0.773, 1.311)
17	1.401	1.465	–0.997	2.361	(1.619, 1.306, 0.981)
7	1.541	1.539	–1.067	2.371	(1.089, 0.783, 0.156)
8	0.058	0.392	0.264	2.399	(0.290, 0.575, 1.293)
1	1.139	1.429	–1.149	2.407	(1.407, 0.830, 0.692)
10	1.723	1.853	–0.801	2.480	(1.086, 0.980, 1.466)
16	0.211	1.415	0.673	2.498	(0.318, 1.138, 1.925)
2	1.515	2	–1.737	2.646	(1.247, 1.201, 0.111)
14	–0.541	0.279	–1.771	2.703	(1.811, 0.714, 0.035)
4	0.772	2	–0.959	2.737	(0.259, 1.241, 0.703)
12	–1.391	1.065	1.640	2.912	(1.685, 1.595, 0.392)
13	0.854	0.672	1.443	3.809	(0.782, 0.629, 0.221)
5	1.902	1.717	1.224	4.500	(1.305, 1.374, 0.710)
11	1.285	0.999	1.391	4.899	(0.615, 0.929, 1.098)
3	1.792	1.243	0.826	8.909	(0.286, 0.013, 1.185)
18	0.466	0.005	1.474	9.369	(0.818, 0.340, 1.791)
15	–1.921	–1.351	0.711	10.888	(1.839, 0.695, 1.204)
9	1.524	0.73	–1.093	12.152	(1.017, 0.002, 0.601)
19	0.265	–1.667	–1.521	35.406	(1.984, 0.320, 0.953)
20	1.827	–1.990	0.234	52.376	(0.969, 0.218, 0.659)

The assignment of the society members to societies based on normalized Euclidean distance is shown in Table 3.14.

Table 3.14: Assignment of the society members to societies.

Civilization leader (CL)	*Society number*	*Society leaders (SL)*	*Society members*
Solution no. 6	1	Solution No. 6	Solution Nos. 5, 11, 3
	2	Solution No. 17	Solution Nos. 1, 4, 9
	3	Solution No. 7	Solution Nos. 10, 2
	4	Solution No. 8	Solution Nos. 16, 14, 12, 13, 18, 15, 19, 20

Step 5: *Update the velocities and position of each member in society*

The velocity and position of each member is then updated using following equations:

For civilization leader (CL):

$$V_{CL\_i+1} = w \times V_{CL\_i} + c_1 \times r_1 \times (pBest_{CL\_i} - X_{CL\_i}) \quad (3.34)$$

$$X_{CL\_i+1} = X_{CL\_i+1} + V_{CL\_i+1} \quad (3.35)$$

For society leader (SL):

$$V_{SL\_i+1} = w \times V_{SL\_i} + c_1 \times r_1 \times (pBest_{SL\_i} - X_{SL\_i}) + c_2 \times r_2 \times (X_{CL\_i} - X_{SL\_i}) \quad (3.36)$$

$$X_{SL\_i+1} = X_{SL\_i+1} + V_{SL\_i+1} \quad (3.37)$$

For society member (SM):

$$V_{SM\_i+1} = w \times V_{SM\_i} + c_1 \times r_1 \times (pBest_{SM\_i} - X_{SM\_i}) + c_2 \times r_2 \times (X_{SM\_i} - X_{SL\_i}) \quad (3.38)$$

$$X_{SM\_i+1} = X_{SM\_i+1} + V_{SM\_i+1} \quad (3.39)$$

Step 6: *Repeat steps 2 to 5 till convergence*

Immunity enhanced PSO

In Immunity-enhanced particle swarm optimization IEPSO (Kang *et al.*, 2013), a population of particles is sampled randomly in the feasible space. The population of particles is used to execute PSO or its variants having the updated values of position and velocity. After that, it executes receptor editing operator also known as non- uniform mutation according to a certain probability (p_r), and vaccination operator according to probability (p_v). The new generation is obtained by the selection operator after the flying of particles and two immune operators receptor editing and vaccination.

Neighborhood search PSO (NSPSO)

Due to the intrinsic randomness, PSO suffers from premature convergence when solving complex multimodal problems. Sometimes, the sub-optima are near to the global optimum and the neighbourhoods of trapped particles may contain the global optimum. For this case, searching the neighbours of particles is beneficial for finding better solutions. Based on this idea, some neighbourhood search strategies have been successfully applied to various algorithms. Wang *et al.* (2011) proposed a new PSO algorithm called PSO with neighbourhood search strategies (NSPSO), which utilizes one local and two global neighbourhood search strategies. The NSPSO includes two operations. First, for each particle, three trial particles are generated by the above neighbourhood search strategies, respectively. Second, the best one among the three trial particles and the current particle is chosen as the new current particle.

3.6.4. *Shuffled frog leaping algorithm*

The SFL algorithm proposed by Eussuff *et al.* (2006) combines the benefits of the genetic-based memetic algorithm and the social behavior-based PSO algorithm. It is a metaheuristic optimization method that mimics the memetic evolution of a group of frogs when seeking for the location that has the maximum amount of available food. In the SFL algorithm, the population consists of a set of frogs (solutions) that is partitioned into subsets referred to as memeplexes. The different memeplexes are considered as different cultures of frogs, each performing a local search. Within each memeplex, the individual frogs hold ideas, that can be influenced by the ideas of other frogs, and evolve through a process of memetic evolution. After a defined number of memetic evolution steps, ideas are passed among memeplexes in a shuffling process. The local search and the shuffling processes continue until defined convergence criteria are satisfied. The detailed working of SFL algorithm and its applications are discussed in the chapter 5.

4

Artificial *Bee* Colony Algorithm

4.1. Introduction

Artificial bee colony algorithm is developed by Karaboga and Basturk (2008) to model the intelligent behaviors of honey-bee swarms. The honeybee swarms consists of three essential components: food sources, employed foragers and unemployed foragers. It defines two leading modes of the behavior: recruitment to a nectar source and abandonment of a source. Fig. 4.1 shows behavior of honeybee foraging for nectar.

(i) *Food sources*: The value of a food source depends on many factors, such as its proximity to the nest, richness or concentration of energy and the ease of extracting this energy. For the simplicity, the "profitability" of a food source can be represented with a single quantity. As shown in Fig. 4.1 there are two discovered food sources: A and B.

(ii) *Employed foragers (EF)*: They are associated with a particular food source, which they are currently exploiting or are "employed" at. They carry with them information about this particular source, its distance and direction from the nest and the profitability of the source and share this information with a certain probability.

(iii) *Unemployed foragers (UF)*: Look for a food source to exploit. There are two types of unemployed foragers scouts (S) searching the environment surrounding the nest for new food sources and onlookers waiting in the nest and finding a food source through the information shared by employed foragers *i.e.,* recruit (R).

The exchange of information among bees is the most important occurrence in the formation of collective knowledge. While examining the entire hive, it is possible to distinguish some parts that commonly exist in all hives. The most important part of the hive with respect to exchanging information is the dancing area. Communication among bees related to the quality of food sources occurs in the dancing area as shown in Fig. 4.1. The related dance is called waggle dance. Since information about all the current rich sources is available to an onlooker on the dance floor, she probably could watch numerous dances and choose to employ herself at the most profitable source. There is a greater probability of onlookers choosing more profitable sources since more information is circulating about the more profitable sources. Employed foragers share their information with a probability, which is proportional to the profitability of the food source and sharing of this information through waggle dancing is longer in duration. Hence, the recruitment is proportional to profitability of a food source.

The performance of artificial bee colony (ABC) algorithm in terms of convergence rate and accuracy of the solution is found superior over other non-traditional techniques such as genetic algorithm, particle swarm optimization algorithm etc. In many recent applications ABC algorithm combines both, the stochastic selection scheme carried out by onlooker bees and greedy selection scheme used by onlookers and employed bees to update the source position. Also the neighbor source production mechanism in ABC is similar to the mutation process, which is self-adapting. The random selection process carried out by the scout bees maintains diversity in the solution. Flowchart for ABC algorithm is shown in Fig. 4.2.

4.2. Mechanism of Working of Artificial *Bee* Colony (ABC) Algorithm

As mentioned earlier, the key operations in ABC algorithm are: assign the onlooker bees to recruits, determine positions of onlooker and scout bees and update the food source.

The working of artificial bee colony (ABC) algorithm is demonstrated through example:

$$\textit{Minimize } z = x_1^2 - 3x_2 + 10 \text{ such that } 0 \le x_1, x_2 \le 3.$$

Step 1: *Parameter selection*

Algorithm specific parameters such as population size *i.e.,* number of food sources (equal to number of employed bees), number of onlooker

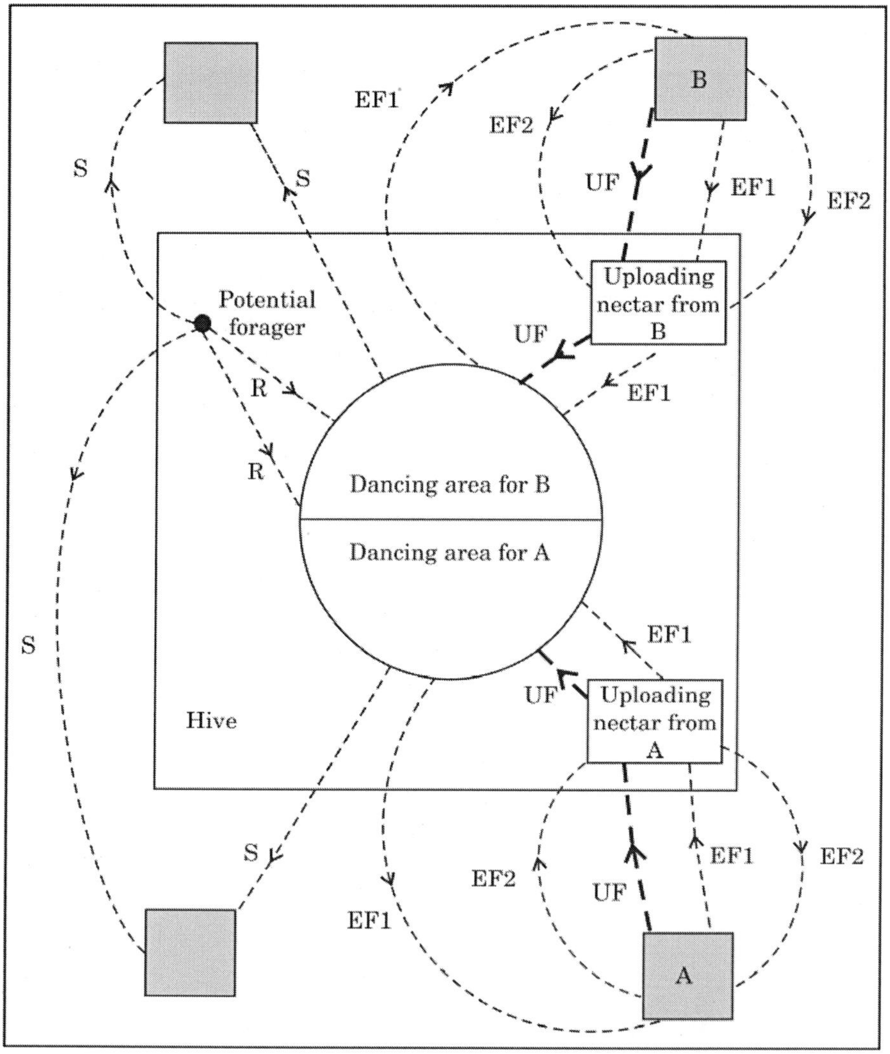

Fig. 4.1: Behavior of honey-bee foraging for nectar (Karaboga and Basturk, 2008)

bees, number of scout bees (5–30% of the colony size) are to be determined and these are as follows:

- Number of food sources: 5
- Number of onlooker bees: 11
- Number of scout bees: 1

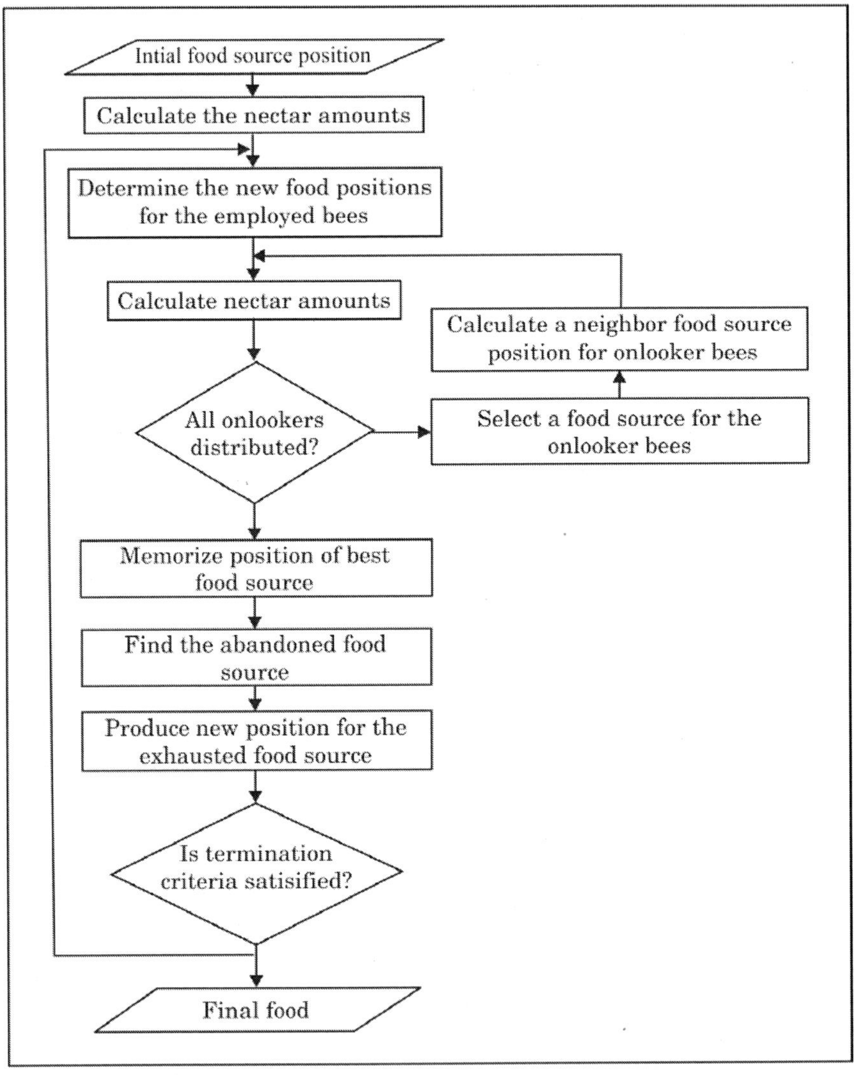

Fig. 4.2: Flowchart for artificial bee colony (ABC) algorithm (Karaboga and Basturk, 2008).

Step 2: ***Calculate the nectar amount (quality) for each food source***

The nectar amount represents the actual fitness value of the solution. Nectar amount of each food source is obtained by evaluating objective function value. Each employed bee is assigned to one food source as shown in Fig. 4.3.

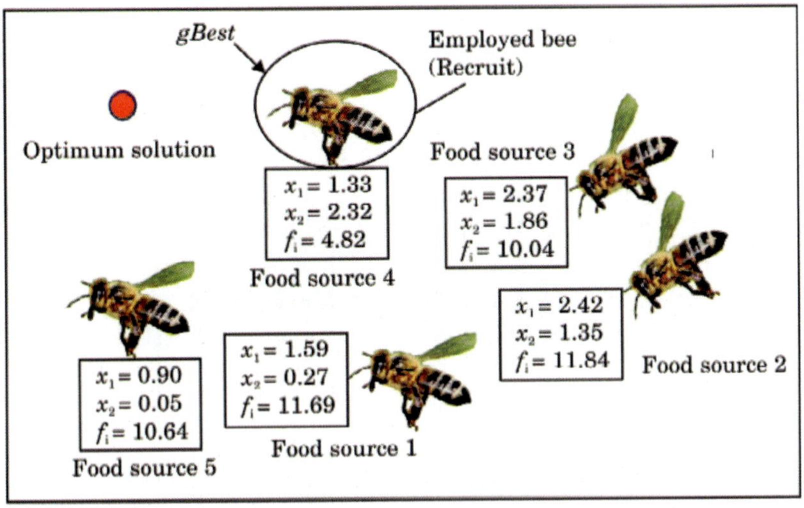

Fig. 4.3: Food source positions and nectar amounts.

Step 3: *Determine the probabilities of assigning onlooker bees to employed bees*

As discussed in section 4.1, the employed bees perform waggle dance to share information about the richness of their food source. The onlooker bees observe this waggle dance and obviously more number of onlooker bees will be attracted towards food source (and thereby corresponding employed bee) which is rich in nectar. For the present example, the number of onlooker bees is 11. Now, the probability *(P_i)* with which the onlooker bee is assigned to employed bee is obtained using Eq. (4.1).

$$P_i = \frac{[\sum_{i=1}^{R}(1/f_i)]^{-1}}{f_i}. \quad (4.1)$$

Where R is number of food sources, f_i is fitness value of food source i.

Using Eq. (4.1), the probabilities of assigning onlooker bees to the food sources (or employed bees) 1 to 5 are 15%, 15%, 17%, 36%, and

16% respectively. It can be seen that, as the food source 4 is richest in nectar (*i.e.,* solution with highest fitness), the probability with which the onlooker bees will be assigned to this food source is also high (36%).

Step 4: *Calculate the number of onlooker bees, which will be sent to food sources*

The number of onlooker bees sent to an employed bee (N) is obtained using Eq. (4.2).

$$N = P_i \times m \tag{4.2}$$

where 'm' is the total number of onlooker bees.

In this case m = 11, hence the number of onlooker bees assigned to the food sources 1 to 5 are: 2, 1, 2, 4, 2 respectively and grouped in circles as shown in Fig. 4.4.

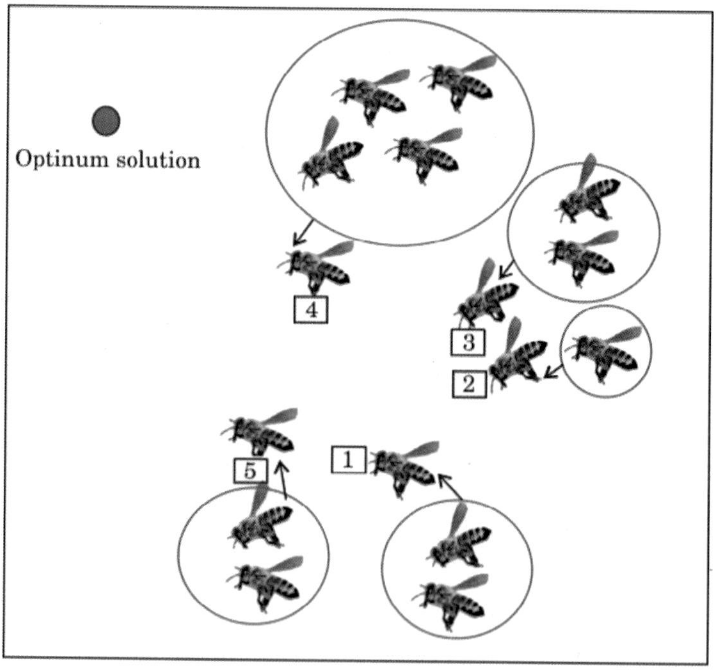

Fig. 4.4: Assignment of onlooker bees to employed bees

Step 5: *Determine new position of each onlooker bee*

Each employed bee is updated (for N times) using Eq. (4.3) to obtain positions of onlooker bees assigned to employed bees (food sources).

$$\theta_i(c+1) = \theta_i(c) \pm \emptyset_i(c) \tag{4.3}$$

Where c is number of generation, ϕ_i(c) is a randomly produced step to find a food source with a more nectar around θ_i. ϕ_i(c) is calculated by taking the difference of the same parts of θ_i (c) and θ_k (c) (k is a randomly produced index) food positions. If the nectar amount ϕ_i (c + 1) at θ_i (c + 1) is higher than at θ_i (c), then the bees go to the hive and share information with others and the position $_i$ (c) of the food source is changed to θ_i (c + 1) otherwise θ_i (c) is kept as it is. If the position θ_i of the food source i cannot be improved through the pre-determined number of trials, then that food source θ_i is abandoned by its employed bee and then it becomes a scout. The scout starts searching for new food source, and after finding the new source, the new position is accepted as θ_i. Table 4.1 shows the distribution of onlooker bees to food source based on the information shared by employed bees (recruits) to the onlooker bees.

Table 4.1: Assignment of onlooker bees to food source.

Food source bee no.	***Onlooker***	***Onlooker bee positions***		***z'***
		x_1	x_2	
1	1	0.80	2.32	3.69
	2	1.85	1.86	7.82
2	1	1.59	2.43	5.24
3	1	1.33	3.00	2.76
	2	2.32	3.00	6.36
4	1	0.28	1.35	6.03
	2	1.07	2.77	2.84
	3	1.76	1.35	9.04
	4	0.28	0.27	9.26
5	1	0.47	1.35	6.16
	2	0.21	2.32	3.10
	Scout	1.16	1.84	5.82

It can be seen from Table 4.1 that, the two onlooker bees are assigned to food source 1, one onlooker bee is assigned to food source 2, two onlooker bees are assigned to food source 3, four onlooker bees are assigned to food sources 4 and two onlooker bees are assigned to food source 5. Position of one scout bee randomly searching food source is also shown. Now the onlooker bee at best position out of onlooker bees assigned to each food source is compared with existing employed bee position as shown in Table 4.2.

Table 4.2: Updated positions of food sources.

Food source	*Existing employed bee position*			*Onlooker bee at best position*			*Updated food source positions*		
	x_1	x_2	z	x_1	x_2	z	x_{1new}	x_{2new}	z_{new}
1	1.59	0.27	11.69	0.80	2.32	3.69	0.80	2.32	3.69
2	2.43	1.35	11.84	1.59	2.43	5.24	1.59	2.43	5.24
3	2.37	1.86	10.04	1.33	3.00	2.76	1.33	3.00	2.76
4	1.33	2.32	4.82	1.07	2.77	2.84	1.07	2.77	2.84
5	0.90	0.05	10.64	0.21	2.32	3.10	0.21	2.32	3.10

As shown in Table 4.2, positions of all food sources are updated. It means all onlooker bees are successful to identify the better food source in the vicinity of the food sources to which they were assigned. It is observed from Table 4.2 that the average fitness value of initial population is 9.806, whereas average fitness value of updated population is 3.526 showing an improvement of 2.7 times over initial population. Also, the fitness values of best solutions in the initial and updated population are 4.82 and 2.76 respectively. Thus, the updated population is better not only with respect to average performance but also with respect to best value. The updated food source positions are shown in Fig. 4.5.

Step 6: *Evaluate the best solution*

Position of the best onlooker bee is identified for each food source. The global best (*gBest*) of the honeybee swarm in each generation is obtained and it may replace the global best at previous generation if it has better fitness value. For initial population, the global best solution is at food source 4 with x_1 = 1.33, x_2 = 2.32, and fitness value = 4.8, while updated population has global best at food source 3 with x_1 = 1.33, x_2 = 3.00, and fitness value = 2.76. Thus, the global best for updated solutions is better than that of previous generation and hence the updated global best replaces previous global best and the new global best is at food source 3.

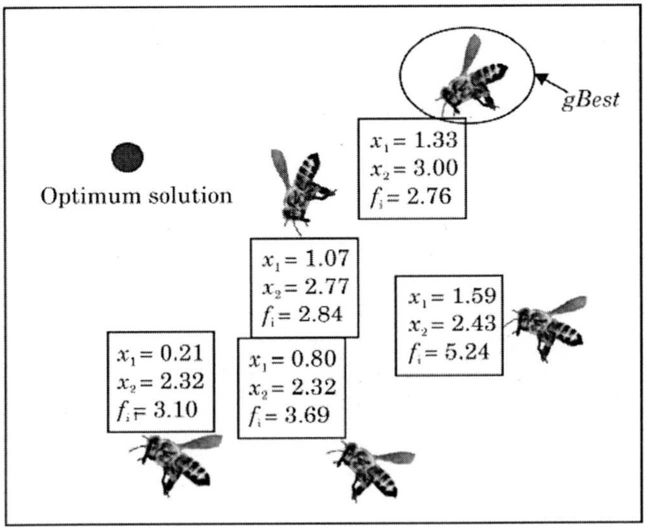

Fig. 4.5: Updated positions of food sources

Step 7: *Update the scout bee*

The worst employed bees, as many as the number of scout bees in the population, are respectively compared with the scout solutions. If the scout solution is better than employed solution, employed solution is replaced with scout solution. Else employed solution is transferred to the next generation without any change. In the present case, scout bee solution (generated randomly) is: x_1 = 1.16, x_2 = 1.84 with fitness value = 5.82. This solution is compared with the worst solution in the updated population which is x_1 = 1.59, x_2 = 2.43 with fitness value = 5.24. The scout solution is thus inferior to worst solution in population and hence the worst solution in updated population will not be replaced by scout solution.

The applications of ABC algorithm are demonstrated through case studies discussed in sections 4.3 and 4.4.

4.3. Design Optimization of Screw Conveyer System for Handling Carbon Black Powder

The problem considered in this case study is an industrial problem employing pyrolysis process. The pyrolysis process is a technique for recycling used tires in which the whole or shredded tires are heated and softened in an oxygen-free reactor vessel, breaking down the rubber polymers into small molecules. These small molecules vaporize and exit from the reactor in the form of fuel oil. The main product of the pyrolysis process is thus a fuel oil and the by-products are carbon black and steel wires. The carbon black powder is collected in the pit and then transferred to ground manually or by means of conveyer. The manual transfer of this free flow powder suffers from several drawbacks such as uncontrolled dispersion of powder, operator fatigue, high labor cost and high cycle time. To minimize these issues screw conveyor as shown in Fig. 4.6 can be a better choice. Screw conveyors are having following advantages over other conveyor systems:

1. They are capable of handling a great variety of bulk materials from sluggish to free-flowing.
2. They can have multiple inlet and discharge points. Bulk materials can be conveyed and distributed to various locations as required. Slide gates or valves can be added to control the flow into and out of a screw conveyor.
3. They are very compact and adaptable to congested locations. Screw conveyors do not have a return similar to a belt or drag conveyor.
4. Screw conveyors are totally enclosed to contain the product and prevent spillage. They can be utilized in the horizontal, vertical or any inclined position depending upon the characteristics of the product being conveyed.
5. Screw conveyors can provide an air lock between upstream and downstream equipment.

However, the effectiveness of the system could be realized through maximizing the material handling rate (Q) and minimizing the cost of equipment (C) These two aspects depends mainly on three factors: (i) Nominal diameter of screw, D (ii) Rotational speed of the screw, N and (iii) Inclination angle of Screw Conveyor, a.

The relationship of these three factors with performance measures *i.e.,* material handling rate (Q) and cost of conveyor (C) can be expressed using following equations:

The approximate theoretical model of material handling rate of a screw conveyor is given as:

$$Q = 60.\ \frac{\pi}{4} D^2\ S.N.\psi.\rho.m\ \text{(kg/hr)} \tag{4.4}$$

Where, ρ = Bulk density of the material (kg/m^3)

D = Nominal diameter of screw in (m)

S = Screw pitch in (m)

N = Rotational speed of screw (rpm)

ψ = Loading factor

m = Factor to take into account the inclination of the conveyor =

$$-0.00355 \propto^2 + 0.0246 \propto + 0.33$$

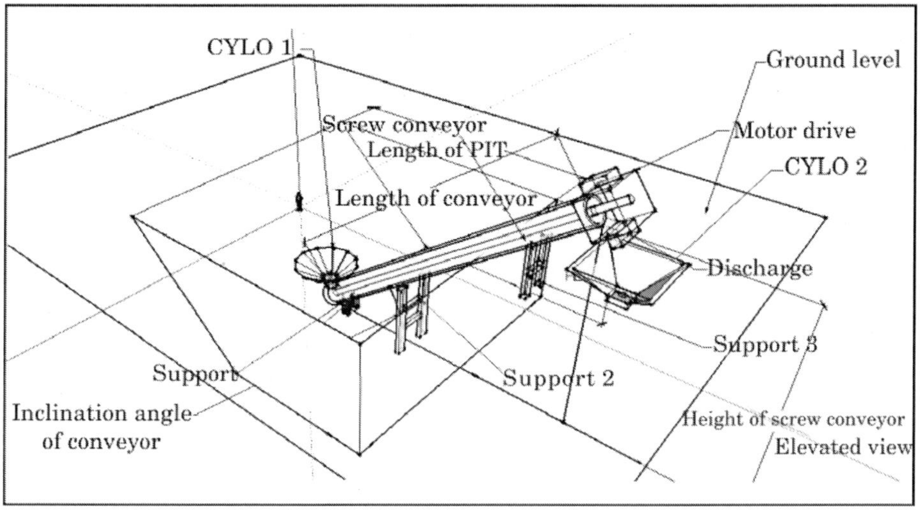

Fig. 4.6: Screw conveyor system for handling carbon black powder from pit to ground.

The screw conveyor has main four parts: trough (a long container carrying screw), screw shaft, screw blades and motor. The cost of the conveyor (C) is mainly due to material costs of these four parts as given below.

$$C = C_{through} + C_{shaft} + C_{blades} + C_{motor}\ \text{(Rs.)} \tag{4.5}$$

Where,

$C_{through}$ is a material cost of a trough and is calculated as:

$$C_{through} = \frac{\pi}{4} \cdot [d_o^2 - D^2] \cdot \frac{H}{\sin\alpha} \cdot \rho_t \cdot C_t \qquad (4.6)$$

Where, d_o = Outer diameter of trough (m), D = Nominal screw diameter (m), H is height at which the material is to be transported (m), a = Inclination Angle of Screw Conveyor (°), ρ_t

= density of trough material (Kg/m³), and C_t cost of trough material per Kg (Rs)

The relation between the height (H), inclination angle of screw (a) and length of a screw conveyor (L) is shown in Fig. 4.7.

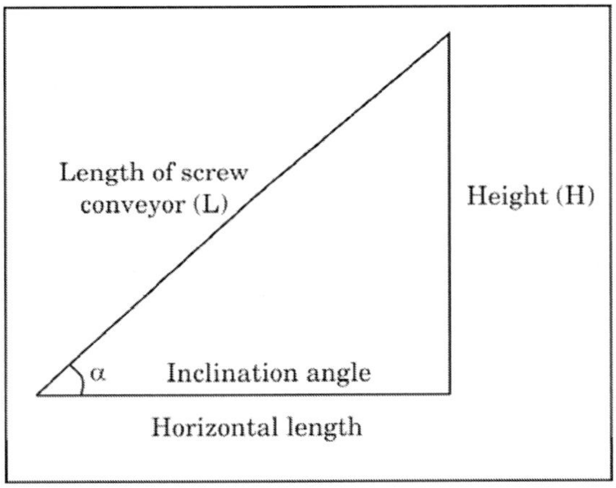

Fig. 4.7: Relation between the height (H), inclination angle of screw (a), and length of a screw conveyor (L).

$$C_{shaft} = \text{Cost of screw shaft material} = \frac{\pi}{4} \cdot d_s \cdot \frac{H}{\sin\alpha} \cdot \rho_s \cdot C_s \text{ (Rs.)} \qquad (4.7)$$

Where, . d_s = screw shaft diameter, r_s = density of screw shaft material (Kg/m³), C_s is the cost of screw shaft material/Kg (Rs)

$$C_{blades} = \text{Cost of screw blade material} = Z.\ \frac{\pi}{4} \cdot [D - \gamma]^2 \cdot t \cdot \rho_b \cdot C_b \text{ (Rs)} \qquad (4.8)$$

Where, Z = Number of helical blades, g = screw clearance, t = thickness of helical blade (m), ρ_b = density of blade material (Kg/m^3), C_b is the cost of blade material/Kg (Rs) Cost of motor (C_{motor}) is a function of its power (P). The relation between the cost and power of the motor is obtained using linear regression for practically known data as shown in Table 4.3.

Table 4.3: Cost of motor as function of its power.

P (hp)	1	3	5	7.5
C_{motor} (Rs)	3000	5500	8000	10000

The regression equation for the above data is:

$$C_{\text{motor}} = 2144.2 + 1086.25.P \tag{4.9}$$

The power P can be calculated as:

$$P = \frac{2\pi NT}{60} \times 746\ (hp) \tag{4.10}$$

Where, T = Torque required to raise the load (N–m) = W. tan $(\theta+\phi)$. $(d_m/2)$.

W = Load carried by conveyor (N) = Maximum volume capacity of material in the conveyor × Density of material × % loading = $\frac{\pi}{4}[D_o^2 - D_i^2].\frac{H}{\sin\alpha}.\rho.(0.4)\times 9.81$. θ is angle of helix = $\frac{pitch\ (k)}{\pi d_m}$, ϕ = Friction angle between the material to handle and the screw surface.

For the present case study, the known parameters are: Ψ = Loading factor = 0.4 (40%), H= 4.72 m, t_t = 0.002 m, $d_0 = \rho$ = 363 kg/m^3, $\rho_t = \rho_s = \rho_b$ = 7850 kg/m^3, $C_t = C_s = C_b$ = 50 Rs/ kg, d_s = 0.0325 m, t_b = 5 mm, γ= 4 mm, θ = 17.65°, Φ = Friction angle = 5.71°, $S = D$.

Substituting these values in Eq. (4.5) and Eq. (4.4), the optimization model is:

***Objective function*:** Objective is to design the screw conveyor configuration for minimum cost.

Minimize cost:

$$C = \left\{\frac{1}{sin\alpha}\left[\ 229021.5D + \frac{0.1164}{D} + 10023.73 + (380D^2 - 0.98 \times D \times N\right]\right\}$$

$$+\ 1923 \qquad (4.11)$$

Constraint: The constraint is to ensure that the material transfer rate should not be less than 2000 Kg/Hr. It can be expressed as:

$$Q = 6842.38 \times D^3 \times N\ (-\ 3.55 \times 10^{-4} \propto^2 + 0.024 \propto + 0.33) \geq 2000 \qquad (4.12)$$

Variables bounds: $30° \leq a \leq 45°$; $0.2286 \leq D \leq 0.3086$ m; $15 \leq N \leq 25$

Now, the ABC algorithm is applied as per the steps discussed in section 2.

The combined objective function is thus formulated as:

$$Min\ z = C + penalty \times (Q - 2000) \qquad (4.13)$$

Static penalty is applied if the term $(Q - 2000) < 0$, else, penalty is zero.

Step 1: ***Parameter selection***

- Number of food sources: 10
- Number of onlooker bees: 21
- Number of scout bees: 2

Step 2: ***Calculate the nectar amount (quality) for each food source***

The nectar amount represents the actual fitness value of the solution. Nectar amount of each food source is obtained by evaluating objective functions values using Eq. (4.13). These values for 10 food sources (initial population) are shown in Table 4.4.

Step 3: ***Determine the probabilities of assigning onlooker bees to employed bees***

The probability (P_i) with which the onlooker bee is assigned to employed bee (or the food source) is obtained as shown in Table 4.5.

Table 4.4: Nectar amount of each food source.

Food source	*a (°)*	*D (m)*	*N (rpm)*	*C (Rs)*	*Q (kg / hr)*	*z*
1	42.64	0.2556	17.93	103305.59	1502.48	227685.23
2	34.99	0.2349	22.92	113381.01	1535.98	229385.53
3	30.18	0.3045	16.00	160921.39	2315.37	160921.39
4 (*gBest*)	38.51	0.2870	21.10	123867.97	2561.81	123867.97
5	36.91	0.2873	22.29	128512.87	2728.58	128512.87
6	31.23	0.3028	19.01	155381.56	2714.94	155381.56
7	42.80	0.2566	19.31	103343.32	1634.93	194611.71
8	39.88	0.2351	23.01	101696.66	1527.02	219942.45
9	35.61	0.2744	23.72	127389.46	2535.34	127389.46
10	36.71	0.2777	23.81	125409.77	2633.51	125409.77

Table 4.5: Probabilities of assigning onlooker bees to food source (onlooker bee).

Food source	*1*	*2*	*3*	*4*	*5*	*6*	*7*	*8*	*9*	*10*
Probability of assigning onlooker bees to food source	0.07	0.07	0.10	0.13	0.12	0.10	0.08	0.07	0.13	0.13

Step 4: *Calculate the number of onlooker bees, which will be sent to food sources*

The number of onlooker bees sent to an employed bee (N) obtained using Eq. (11) as shown in Table 4.6.

Table 4.6: Number of onlooker bees assigned to food source

Food source	*1*	*2*	*3*	*4*	*5*	*6*	*7*	*8*	*9*	*10*
Number of onlooker bees assigned to food sources	1	1	2	3	3	2	2	1	3	3

Step 5: *Determine new position of each onlooker bee*

Each employed bee is updated (for N times) to obtain positions of onlooker bees assigned to it. Table 4.7 shows the positions of all onlooker bees and scout bees.

Onlooker bees which discovered best food sources in their group are marked bold in Table 4.7. These onlooker bees are compared with their corresponding employed bees. The positions of the existing employed bees and onlooker bees at best positions are shown in Table 4.8. If the food source discovered by the best onlooker bee is better in nectar than that of corresponding employed bee, the onlooker bee will replace the employed bee and the existing employed bee position will be abandoned.

Table 4.7: Assignment of onlooker bees to food source.

Food source	*Onlooker bee No.*	*Onlooker bee positions*			*C (Rs)*	*Q (kg/hr)*	*z'*
		a (°)	*D (m)*	*N (rpm)*			
1	1	41.72	0.3000	16.66	120143.52	2250.29	**120143.52**
2	1	36.71	0.2286	23.24	106454.18	1433.48	**248083.44**
3	1	30.40	0.2951	16.68	155594.58	2197.52	**155594.58**
	2	32.89	0.2891	15.00	142547.57	1872.59	174400.75
4	1	37.53	0.2625	21.55	117315.39	2009.84	**117315.39**
	2	35.83	0.3048	20.61	138650.54	3017.88	138650.54
	3	30.00	0.2883	19.04	154363.50	2336.30	154363.50
5	1	44.10	0.3048	21.59	116955.00	3029.84	**116955.00**
	2	39.01	0.2482	21.81	108331.54	1709.87	180863.15
	3	38.80	0.2511	22.45	109900.63	1824.36	153810.86
6	1	32.23	0.3026	18.70	151031.09	2673.38	151031.09
	2	31.13	0.2882	23.82	149416.02	2934.34	**149416.02**
7	1	34.75	0.2913	19.37	136839.27	2476.61	**136839.27**
	2	42.70	0.2543	20.76	102767.30	1712.98	174521.10
8	1	40.34	0.2337	23.06	100237.46	1499.27	**225420.57**
9	1	39.69	0.2689	20.61	114275.68	2048.32	114275.68
	2	40.46	0.2685	24.82	112385.40	2445.43	**112385.40**
	3	33.33	0.2760	23.71	135514.09	2575.41	135514.09
10	1	35.78	0.2781	23.46	128320.06	2609.63	128320.06
	2	39.16	0.2741	21.08	117464.06	2225.57	**117464.06**
	3	39.97	0.2439	23.11	104663.13	1711.45	176800.59
	Scout 1	37.02	0.29	15.21	129802.23	1951.13	142018.76
	Scout 2	34.91	0.23	22.47	112054.25	1433.30	253729.90

It can be seen from Table 4.8 that, onlooker bees are successful to search eight new food sources which are better in nectar. Only the food sources 2 and 8 are not able to update. These onlooker bees are now recruits. The updated food source positions are shown in Table 4.9.

Step 6: ***Evaluate the best solution***

For initial population the global best solution is at food source 4 with $a = 38.51°$, $D = 0.2870$ m, and $N = 21.10$ rpm with fitness value = 123867.97, while for updated population the global best is at food source

9 with a = 35.61°, D = 0.2744 m, and N = 23.72 rpm with fitness value = 112385.40. Thus, the global best for updated solutions is better than that of previous generation and hence the updated global best replaces previous global best and new global best is at food source 9.

Table 4.8: Food sources identified by onlooker bees.

Food source	*Existing employed bee positon*				*Onlooker bee at best position*			
	a (°)	*D* (m)	*N* (rpm)	*z*	*a* (°)	*D* (m)	*N* (rpm)	*z*
1	42.64	0.2556	17.93	227685.23	41.72	0.3000	16.66	**120143.52**
2	34.99	0.2349	22.92	**229385.53**	36.71	0.2286	23.24	248083.44
3	30.18	0.3045	16.00	160921.39	30.40	0.2951	16.68	**155594.58**
4	38.51	0.2870	21.10	123867.97	37.53	0.2625	21.55	**117315.39**
5	36.91	0.2873	22.29	128512.87	44.10	0.3048	21.59	**116955.00**
6	31.23	0.3028	19.01	155381.56	31.13	0.2882	23.82	**149416.02**
7	42.80	0.2566	19.31	194611.71	34.75	0.2913	19.37	**136839.27**
8	39.88	0.2351	23.01	**219942.45**	40.34	0.2337	23.06	225420.57
9	35.61	0.2744	23.72	127389.46	40.46	0.2685	24.82	**112385.40**
10	36.71	0.2777	23.81	125409.77	39.16	0.2741	21.08	**117464.06**

Table 4.9: Updated food source positions:

Food source	*Existing employed bee position*			
	a(°)	*D* (m)	*N* (rpm)	*z*
1	41.72	0.3000	16.66	120143.52
2	34.99	0.2349	22.92	229385.53
3	30.40	0.2951	16.68	155594.58
4	37.53	0.2625	21.55	117315.39
5	44.10	0.3048	21.59	116955.00
6	31.13	0.2882	23.82	149416.02
7	34.75	0.2913	19.37	136839.27
8	39.88	0.2351	23.01	219942.45
9 (*gBest*)	40.46	0.2685	24.82	112385.40
10	39.16	0.2741	21.08	117464.06

Step 7: *Update the scout bee*

As discussed in section 4.2, the worst food sources as many as the number of scout bees in the population, are respectively compared with the scout solutions. As two scout bees are considered for this case study, two worst food sources in the updated population are compared with the positions of these two scout bees as shown in Table 4.10.

Table 4.10: Worst food sources compared with scout bees positions

Food source	*Updated positions worst employed bees*				*Scout bee*	*Scout bees positions*			
	a (°)	*D* (m)	*N* (rpm)	*z*		*a* (°)	*D* (m)	*N* (rpm)	*z*
2	34.99	0.2349	22.92	229385.53	1	37.02	0.29	15.21	142018.76
8	39.88	0.2351	23.01	219942.45	2	34.91	0.23	22.47	253729.90

It can be seen from Table 4.10, that scout bee 1 is at better position than updated food source 2; hence, food source 2 is replaced by scout bee 1. The final updated food source positions are as shown in Table 4.11.

Table 4.11: Final food positions after first generation.

Food source	*Existing employed bee position*			
	a (°)	*D* (m)	*N* (rpm)	*z*
1	41.72	0.3000	16.66	120143.52
2	37.02	0.29	15.21	142018.76
3	30.40	0.2951	16.68	155594.58
4	37.53	0.2625	21.55	117315.39
5	44.10	0.3048	21.59	116955.00
6	31.13	0.2882	23.82	149416.02
7	34.75	0.2913	19.37	136839.27
8	39.88	0.2351	23.01	219942.45
9 (*gBest*)	40.46	0.2685	24.82	112385.40
10	39.16	0.2741	21.08	117464.06

The steps 1 to 7 are repeated till the convergence.

4.4. Parametric Optimization of Hard Chrome Electro-Plating Process for Uniform Coating Thickness and Improved Hardness

Electroplating also known as electro-chemical plating is an electrolytic process. In this process metal ions in an electrolyte solution are deposited onto a cathode. In the electrolytic process, anode is generally made of metal being plated so it serves as source of coating metal. Work-piece where coating is to be applied made as cathode. Direct current from an external power source is passed through electrolyte solution. The electrolyte is an aqueous solution of acids, bases or salts. Electrolyte conducts electric current by the movement of plate metal ions in solution. Electroplating may have plating of different metals such as:

- ***Zinc coating*:** Zinc plated steel products are fasteners wire goods, electric switch boxes and sheet metal parts. Zinc coating provide high resistant to corrosion.
- ***Nickel plating*:** Nickel plating is preferred for corrosion resistant purpose and decorative purpose. Metals like brass, steel, zinc etc. serve as substrate material.
- ***Gold plating*:** It is a method of depositing a thin layer of gold on the substrate of other metal. Most often substrate is silver or copper. It provides corrosion resistant and highly electrically conductive layer which is used in electrical connectors and printed circuit boards. Some metal may also be gold plated to improve their aesthetic value for ornamental purpose.
- ***Chrome plating*:** The hard chromium plating process is one of the most effective ways of protecting the base material against a hostile environment or improving the surface properties of the base material.
- ***Tin plating*:** Tin plating is used for ferrous and non-ferrous metallic surfaces. Tin is a nontoxic, ductile and corrosion resistance material so it is widely used in food packaging. Electroplating of tin is also used in electronics industry because of its ability to protect the base metal from oxidation. It is used to protect solderability.
- ***Alloy plating*:** Two different metals are such as Nickel and Cobalt are plated simultaneously in case of alloy plating.

The selection of coating of appropriate metal depends on many criteria like breadth of electrode, length of electrode, distance between electrodes, electrolytic concentration, current value, voltage value, layer thickness, surface cleaning, corrosion resistance, dullness, roughness, environmental factor, adhesions, cohesion, hardness of plating, coefficient of friction, surface tension, deposition rate, deposition time, wear resistance, part geometry, part irregularity, friction of plating parts, heat resistance, color of plating, impurities impingement, cost of process, etc. These parameters will decide the life, durability, capability, gradability and operating economy of the electroplating. The primary objective is to control the wear or erosion so that the user adaptability of product does not exceed the certain level (Kumar and Agrawal, 2009).

Of which, if we consider five important criteria such as cost, adhesion, aesthetics, coating thickness possible and hardness, it is well known fact that hard chrome plating is always a better choice with respect to all criteria except aesthetics. Hence, electro deposition of hard chrome deposits is widely recognized and being extensively used for many years in applications involving sliding or revolving parts like rotating shafts, parts of moving components like pistons, valves, rolls, compressors, etc. It enhances the life of a variety of metal parts that experience wear due to abrasion, friction and corrosion. The main advantage of this process is that it is relatively low temperature process and hence can be used to impart hard surface to any component without actually deforming it. These electrolytic coatings with high hardness/strength and wear/corrosion resistance are widely used in the industry to prevent damage to substrate materials. However, this process suffers from few drawbacks such as non-uniform thickness of coating and insufficient hardness. Non-uniform coating thickness is not desirable where close tolerance required, it also increases the wastage of plating material and increases the machining cost for grinding. Insufficient hardness results into plastic deformation due to indentation or abrasion for a given loading condition.

Hard Chromium electroplating process consists of four parts: 1) the external circuit, consisting of a source of direct current (DC), a means of conveying this current to the plating tank; 2) the negative electrodes or cathodes, which are the material to be plated, along with means of positioning the work in the plating solution so that contact is made with the current source; 3) the plating solution; 4) the positive electrodes, the anodes, usually of the metal being plated but sometimes of a conducting material which serves merely to complete the circuit. For the present case study, a mild steel bar of size ϕ 24 × 300 is considered to study the effect of process parameters of electroplating process such as Current Density (x_1), Distance between electrodes (x_2), Temperature(x_3), and Voltage (x_4) on surface hardness and uniformity of coating thickness in terms of deviation from desired coating thickness of 30 μm.

To establish the relationship between the process parameters and the objectives a mathematical model can be developed based on response surface methodology (RSM). Using RSM, sufficient data is collected through designed experimentation. An experiment is designed with 2^k (where, k is number of variables, in the case study k = 4) factorial with central composite-second order ratable design is used. This consist of number of corner points = 16, number of axial points = 8, and a center point at zero level = 4. The axial points are located in a coded test condition space through parameter '*a*'. For the design to remain rotatable, '*a*' is determined as $(2^k)^{1/4} = 2$. Thus the coded level for the axial points is at 2. The center point is repeated four times to estimate the pure error. A second order polynomial response is then fitted to the data into Eq. (3.29)

Where '*y*' is the response and the x_i are coded levels of *k* quantitative variables as given in Table 4.12.

Table 4.12: Coded values of process parameters

Parameters	*Unit*	*Symbol*	*Coded levels*				
			–2	*–1*	*0*	*1*	*2*
Current density	A/inch^2	x_1	2	2.5	3	3.5	4
Distance between electrodes	inch	x_2	5	5.5	6	6.5	7
Temperature	°C	x_3	52	54	56	58	60
Voltage	volts	x_4	6	7	8	9	10

Few factors other than the mentioned above are set at constant values as shown below:

- ***Tank capacity*:** 20000 litres
- ***Concentration*:** *Chromic acid:* 350 gm/litre, *Sulphuric acid:* 4 ml/litre
- ***Density of concentration*:** 28 A/inch^2
- ***pH concentration*:** 4
- ***Time to complete plating operation*:** 30 min.

Using the experimental data, the mathematical formulation of the two objective functions in the form of Eq. (3.29), is as given below (Pawar and Deshmukh, 2016).

$$\textit{Thickness Deviation} = 13.5098 + 0.62728x_1 + 3.0599x_2 - 1.843x_3 -$$

$$6.39075x_4 - 3.55194x_1^2 - 0.99669x_2^2 - 1.28731x_3^2 + 2.77815x_4^2 +$$

$$2.02994x_1 \cdot x_2 - 1.58874x_1 \cdot x_3 - 2.07433x_1 \cdot x_4 - 0.9381x_2 \cdot x_3 -$$

$$1.63954\,x_2 \cdot x_4 + 1.286788\,x_3 \cdot x_4 \tag{4.7}$$

$$\begin{aligned} Hardness = {} & 1218.45 + 27.025\,x_1 + 49.90833\,x_2 - 115.2583\,x_3 - \\ & 22.95\,x_4 - 70.525\,x_1^2 - 28.9875\,x_2^2 - 61.6125\,x_3^2 - 53\,x_4^2 + \\ & 24.8875\,x_1 \cdot x_2 + 40.1625\,x_1 \cdot x_3 + 1.6125\,x_1 \cdot x_4 + \\ & 63.925\,x_2 \cdot x_3 + 29.925\,x_2 \cdot x_4 - 74.975\,x_3 \cdot x_4 \end{aligned} \tag{4.8}$$

The optimization model is thus formulated as:

***Objective function*:** Minimize the coating thickness deviation from desired coating thickness of 30 mm (D_T) as given by Eq. (4.7)

***Constraint*:** The constraint is to ensure that the hardness (H_V) should not be less than 1200 H_V which can be expressed as:

$$H_V - 1200 \geq 0$$

Where, H_V is as given by Eq. (4.8).

***Parameter bounds*:**

$$-2 \leq x_i \leq 2 \quad (i = 1 \text{ to } 4)$$

Now the steps 1 to 7 of ABC algorithm are applied to solve this optimization model.

4.5. Variants of ABC Algorithm

To improve the performance of ABC algorithm, the researchers had proposed several variants of ABC algorithm as discussed below:

4.5.1. *ABC-solution acceptance rule*

In ABC-solution acceptance rule (Alkin and Erdal, 2016), three search mechanisms with different diversification and intensification characteristics are employed. Moreover, three different search mechanism selection probabilities are introduced to control the balance between diversification and intensification. In their proposed approach, a search mechanism is established using the selection probabilities to generate a new neighbor solution from the current one. Additionally, a solution acceptance rule is implemented, in which not only better solutions but also worse solutions may be accepted by using a probability function. A non-linearly decreasing acceptance probability function is employed, thus allowing worse solutions to be more likely accepted in the early phases of the search. Therefore, ABC-solution acceptance algorithm explores the search space more widespread, especially in the early phases of the search process. By using solution acceptance rule and implementing different search mechanisms of contrasting nature, the proposed algorithm balances the trade-off between diversification and intensification efficiently. The main characteristics of this algorithm that differs from the ABC algorithm are:

(i) Three different search mechanisms are employed with varying diversification and intensification abilities. Probabilistic multi-search with predetermined probability values are employed to determine the search mechanism to be used to generate candidate solutions. Therefore, this algorithm explores and exploits the search space efficiently.

(ii) Instead of a greedy selection, a new candidate solution acceptance rule is integrated, where a worse solution may have a chance to be accepted as new solution. This new acceptance rule helps to achieve better diversification performance, specifically in the early phases of the search.

4.5.2. *Hybrid ABC-SA algorithm*

Artificial bee colony (ABC) algorithm and simulated annealing (SA) algorithm are both independently valid approaches with certain strengths and weaknesses. Pawar (2010) proposed new hybrid ABC-SA algorithm to overcome the drawbacks of existing ABC and SA algorithms and to combine the benefits of both the ABC and SA algorithm. ABC algorithm combines both, the stochastic selection scheme carried out by onlooker bees and greedy selection scheme used by onlookers and employed bees to update the source position. Also, the neighbor source production mechanism in ABC is similar to the mutation process, which is self-adapting. The random selection process carried out by the scout bees maintains diversity in the solution. However, ABC algorithm suffers from drawback of inferior local selection strategy and poor local convergence speed. Simulated annealing algorithm has better convergence property if starting temperature is sufficiently high and cooling rate is low. Because of the discrete nature of the function and construction evaluations, the convergence or transition characteristics are not affected by the continuity of differentiability of the function. The convergence is also not influenced by the convexity status of the feasible space. This algorithm can be used to solve mixed-integer, discrete, or continuous problems. However as SA algorithm does not begin with the population of solutions, parallelism cannot be easily exploited in SA (Wang *et al.,* 2005). Thus, the hybrid ABC-SA algorithm combines the advantages of both ABC and SA algorithms. The algorithm begins with allocation of the onlooker bees to the employed bees based on the fitness values of employed bees. In conventional ABC algorithm, the best onlooker bee will replace the employed bee only if it has better fitness value. In hybrid ABC-SA algorithm, a local selection strategy of SA algorithm is applied to select the onlookers' probability related to current temperature. The flowchart for hybrid ABC-SA algorithm is shown in Fig. 4.8.

4.5.3. *Hybrid ABC-Taguchi algorithm*

Yildiz (2013) proposed a Hybrid ABC-Taguchi algorithm to improve the performance of the algorithm in terms of accuracy of solution, convergence rate and number of function evaluations required. The Taguchi's method is a universal approach, which is widely used in robust design. There are three stages to achieve Taguchi's objective: (1) concept design, (2) robust parameter design, and (3) tolerance design. The robust parameter design is used to determine the levels of factors and to minimize the sensitivity of noise. That is, a parameter setting should be determined with the intention that the product response has minimum variation while its mean is close to the desired target. Taguchi's method is based on statistical and sensitivity analysis for determining the optimal setting of parameters to achieve robust performance. The responses at each setting of parameters were treated as a measure that would be indicative of not only the mean of some quality characteristic, but also the variance of that characteristic. The mean and the variance would be combined into a single performance measure known as the signal-to-noise (S/N) ratio. Taguchi's method uses an orthogonal array and analysis of mean to analyze the effects of parameters based on statistical analysis of experiments. To compare performances of parameters, the statistical test known as the analysis of variance (ANOVA) is used. When hybridized with ABC algorithm, the Taguchi method is used determine the optimal levels of the decision parameters and their contributions. If the contribution of a particular

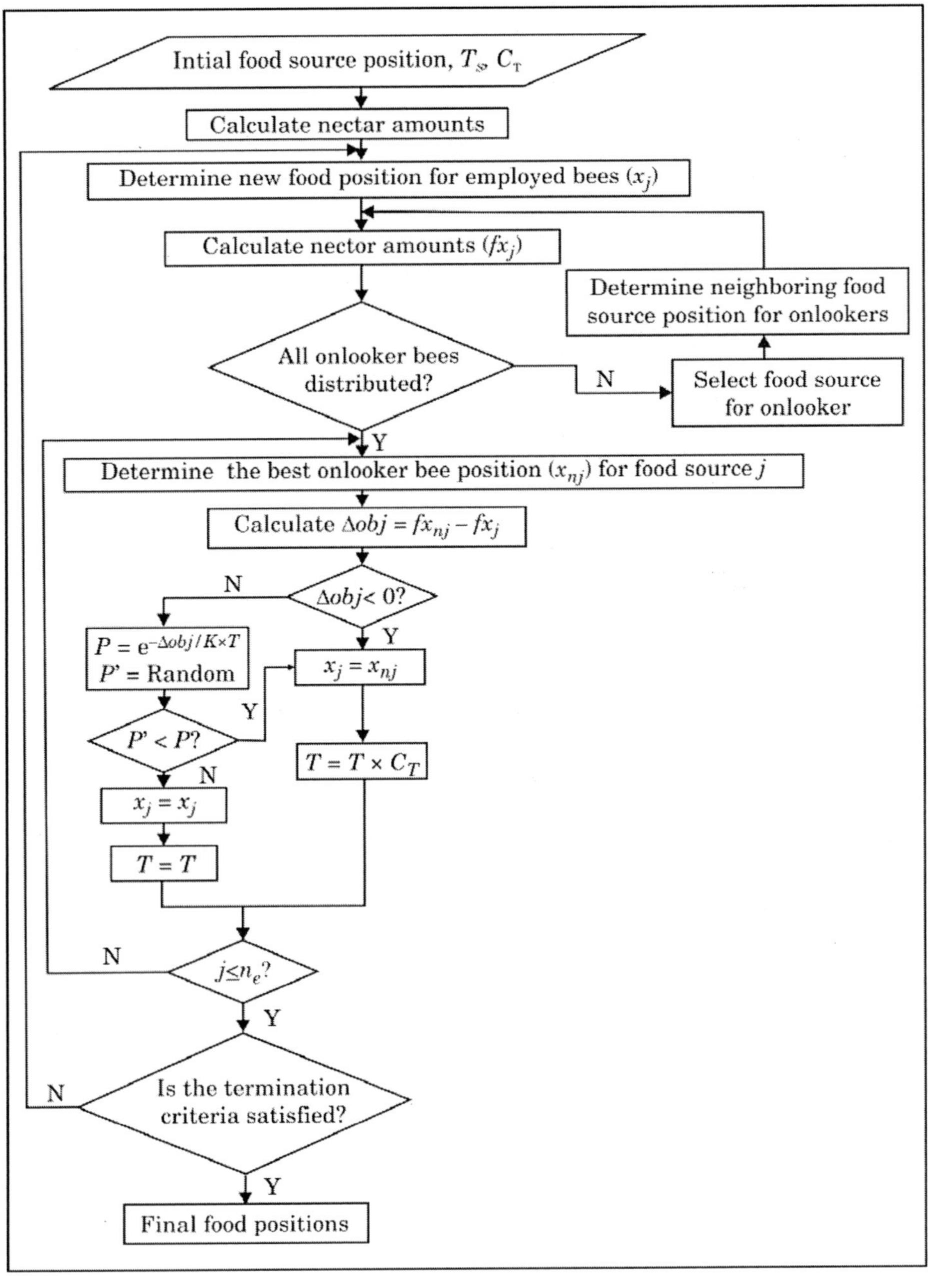

Fig. 4.8: Hybrid ABC-SA algorithm (Pawar, 2010).

variable is significant, then its value at optimum level is used in the ABC algorithm instead of considering entire range of that parameter. This concept is demonstrated through problem formulated in section 4.3, of design optimization of screw conveyer system for handling carbon black powder. The three variables considered in this case study are inclination angle of screw conveyor (a), Nominal screw diameter in meter (D), and Rotational speed of screw (N) with their bounds as:

$$30° \leq a \leq 45°;\ 0.2286 \leq D \leq 0.3086 \text{ m};\ 15 \leq N \leq 25.$$

Now, let us consider 4 levels of each variable as shown in Table below:

Table 4.13: Level chart of variables.

Variable/levels	*1*	*2*	*3*	*4*
a	30	35	40	45
D	0.2286	0.2552	0.2819	0.3086
N	15	18.333	21.667	25

For 3 variables and 4 levels, the L_{16} orthogonal array is constructed as shown in Table 4.14.

Table 4.14: L_{16} Orthogonal array and S/N ratio.

x_1	x_2	x_3	a	D	N	z	*S/N ratio*
1	1	1	30	0.2286	15	397374.69	–111.98
1	2	2	30	0.2252	18.333	357186.35	–111.06
1	3	3	30	0.2819	21.667	151452.21	–103.61
1	4	4	30	0.3086	25	163869.21	–104.29
2	1	2	35	0.2286	18.333	327546.82	–110.31
2	2	1	35	0.2252	15	387848.24	–111.77
2	3	3	35	0.2819	21.667	132275.38	–102.43
2	4	4	35	0.3086	25	143099.70	–103.11
3	1	3	40	0.2286	21.667	268816.34	–108.59
3	2	4	40	0.2252	25	233564.02	–107.37
3	3	1	40	0.2819	15	187226.27	–105.45
3	4	2	40	0.3086	18.333	125546.89	–101.98
4	1	4	45	0.2286	25	221343.28	–106.90
4	2	3	45	0.2252	21.667	283130.92	–109.04
4	3	2	45	0.2819	18.333	105479.35	–100.46
4	4	1	45	0.3086	15	114127.03	–101.15

Table 4.14 shows 16 experiments with coded values of parameters (x_1 to x_3), corresponding actual values of parameters (*a, D, N*), fitness values *z* and S/N ratios. S/N ratios are signal-to-noise ratios and are computed as:

$$\text{S/N ratio} = -10\log\left(\frac{\sum z_i^2}{n}\right) \tag{4.9}$$

Where, i=1 to n are number of replications. For the present case, n=1.

The main issue of experimental analysis is the analysis of variance (ANOVA) which is formed using S/N ratios for the objective function. According to results of ANOVA, appropriate levels of design parameters are found. The result of ANOVA are presented in Table 4.15, the value –107.73 is the average of S/N ratios for level 1 of variable x_1.

Table 4.15: Results of ANOVA.

Levels/ Para- meters	*1*	*2*	*3*	*4*	*SS*	*DOF*	*MS%*	*Contrib- ution*
x_1	−107.73	−106.91	−105.85	−103.39	22.62	2.00	11.31	11
x_2	−109.44	−109.81	−102.99	−102.63	184.17	2.00	92.09	85
x_3	−107.59	−105.95	−105.91	−105.41	8.32	2.00	4.16	4

Similarly other values are computed. For a particular variable, the level having maximum S/N ratio is the optimum level. It can be seen from the Table 4.15, that the optimum level combination is 4-4-4 *i.e.,* 4th level (a = 45°) is most suitable value for variable x_1, 4th level (D=0.3086 m) is most suitable value for variable x_2 and 4th level (N = 25 rpm) is most suitable value for variable x_3. The column *SS* represents the sum of square values. The sample calculation for *SS* value of variable x_1 is shown in Table 4.16.

Table 4.16: Sample calculation for *SS* value of variable x_1.

Levels	*Variable* x_1	*Sum of S/N ratios for level i(*S_i*) B =* $S^2/4$
1	–430.93	46425.17
2	–427.62	45714.72
3	–423.38	44812.66
4	–417.55	43587
	ΣB	180539.53

Levels	*Variable x_1 Sum of S/N ratios for level i(S)*	*$B = S^2/4$*
1	–430.93	46425.17
2	–427.62	45714.72
3	–423.38	44812.66
4	–417.55	43587
	S*B*	180539.53

$$\text{Let } C = \frac{Square\ of\ sum\ of\ all\ \ S/N\ ratios}{16} = 180516.91$$

$$SS = \Sigma B - C = 22.62.$$

Similarly, the *SS* values for other variables are computed. *DOF* = Number of levels –1 = 2. The mean square values (*MS*) are calculated as:

$$MS = \frac{SS}{DOF}$$

The percent contribution of a particular variable is then computed as:

$$\%\ contribution\ \frac{MS}{\sum MS} \times 100$$

As shown in Table the % contribution for variable x_2 is significant (85%) and its optimum level is 4 *i.e.,* D = 0.3086 m. Thus while operating ABC algorithm, value of nominal screw diameter (D) is fixed at 0.3086 m. However, for other variables, the levels can't be fixed since their contributions are weak. The outcome of this exercise is to redefine the bounds of variables as:

$$30° \leq a \leq 45°;\ D = 0.3086\ \text{m};\ 15 \leq N \leq 25.$$

Now the steps of ABC algorithm will be applied to find out the optimum solution.

4.5.4. *ABC with time varying strategy*

ABC algorithm has been shown to be competitive with some other population-based algorithms. However, there is still an insufficiency that ABC is good at exploration but poor at exploitation. To make a proper balance between these two conflictive factors, Qin *et al.* (2015) proposed a novel ABC variant with a time-varying strategy where the ratio between the number of employed bees and the number of onlooker bees varies with time. The linear and nonlinear time-varying strategies can be incorporated into the basic ABC algorithm.

During the process of the employed bees, each food source is assigned with one employed bee. The onlooker bees select the food source based on the fitness values, which is similar to "route wheel selection" in genetic algorithm. Due to this selection scheme, it is assured that the food sources with higher fitness value have more chance to be selected by the onlooker bees. It facilitates improving the quality of these food sources. Thus, the onlooker bees are more concentrated on the exploitation ability than the employed bees. In general, for population-based optimization algorithms, the exploration is typically preferred at the early stages of the search but is required to gradually give ways to exploitation in order to find the optimum solution efficiently. In the ABC algorithm, the population size is fixed. With a large number of employed bees and relatively a small number of onlookers at the early search, it is helpful to move around the search space and enhance the exploration. On the other hand, the small number of the employed bees and the large number of the onlookers allow the individuals to converge to the global optima in the later part of the optimization. In the basic ABC algorithm, the employed bees and onlooker bees are usually equal to each other; that is, the ratio between the number of the employed bees (N_e) and the number of the onlookers (N_o) is 1:1. Considering those concerns, in a time-varying strategy, the ratio between the number of the employed bees and the number of the onlookers varies with time. Linear-decreasing time-varying strategy (LTVS) may contribute to balance the exploration and exploitation during the entire search. A LTVS is proposed as the follows:

$$r_c = (r_{max} - r_{min}) \times \frac{FITc}{FE} \tag{4.10}$$

Where, r_c is ratio of employed bees to onlooker bees, *FITc* is the current number of function evaluations; and *FE* is the total number of function evaluations. The number of employed bees (N_e) are then determined using Eq. (4.11).

$$N_e = round(r_c \times N) \tag{4.11}$$

Where, N = Population size = $N_e + N_o$.

If the number of employed bees is greater than the number of food source, each employed bee is assigned to one food source randomly in the first employed bees, and each of the rest of employed bees is placed at a randomly selected food source. When number of employed bees are smaller than number of food sources, all employed bees randomly select food sources to search.

4.5.5. *ABC with golden section search* (Sharma *et al.*, 2012)

In order to improve the performance of basic ABC, golden section search is introduced in the structure of basic ABC. This process helps in balancing exploration and exploitation. The poor balance may result in weak optimization method which may suffer from trapping in a local optima, stagnation and premature convergence. Introduction of golden section search helps in improving the global convergence and to prevent to stick on a local food source of the ABC. In the proposed algorithm the onlooker bee replaces the food sources in basic ABC using the following Equation:

$$\theta_i(c+1) = \theta_i(c) \pm F_z(x_{ij} - x_{kj}) \ i = k \tag{4.12}$$

Where, F_z is scaling factor depends on the value of a and b discussed below. The generation of a new food sources is seen as an opaque procedure depending on the scale factor F_z. Scaling factor becomes an important aspect that is to be controlled in order to guarantee a high quality food source (solution) that can have an important role in the succeeding generations.

A local search procedure is applied to Fz in order to detect the scale factor value which guarantees a food sources with a high performance. This can be gained by minimizing the function $f(F_z)$ in the decision space [-1, 1]. The meaning of the negative scale factor is the inversion of the search direction. If the search in the negative direction succeeds, the corresponding positive value is assigned to the food sources for the subsequent generations. The Scale factor golden section search applies the Golden section search to the scale factor in order to generate a high quality food sources. This scheme processes in the interval [a = ⁻1, b = 1] and generates two intermediate points:

$F_z^1 = b - \frac{b-a}{\delta}$ and $F_z^2 = a + \frac{b-a}{\delta}$ where, d is the golden number whose value is 1.618.

The evaluated values of $f\,(F_z^1)$ and $f\,(F_z^2)$ are then compared and if < then replaces b elsereplaces a and this procedure is repeated in the new smaller interval [a, b].

5

Shuffled Frog Leaping Algorithm

5.1. Introduction

The shuffled frog leaping (SFL) algorithm is a meta-heuristic optimization method proposed by Eusuff *et al.* (2006) that mimics the memetic evolution of a group of frogs when seeking for the location that has the maximum amount of available food. It is a combination of deterministic and random approaches. The deterministic strategy allows the algorithm to use response surface information effectively to guide the heuristic search as in PSO. The random elements ensure the flexibility and robustness of the search pattern. The search begins with a randomly selected population of frogs covering the entire swamp. The population is partitioned into several parallel communities (memeplexes) that are permitted to evolve independently to search the space in different directions. Within each memeplex, the frogs are infected by other frogs' ideas; hence they experience a memetic evolution. Memetic evolution improves the quality of the meme of an individual and enhances the individual frog's performance towards a goal. To ensure that the infection process is competitive, it is required that frogs with better memes (ideas) contribute more to the development of new ideas than frogs with poor ideas. During the evolution, the frogs may change their memes using the information from the memeplex best or the best of the entire population. Incremental changes in memotype(s) correspond to a leaping step size and the new meme corresponds to the frog's new position. After an individual frog has improved its position, it is returned to the community. The information gained from a change in position is immediately available to be further improved upon. This instantaneous accessibility to new information separates this approach from a GA that requires the entire population to be modified before new insights are available. Again, the visual representation of frogs is used as the analogy here but, in terms of spreading of ideas, the analogy can be teams of inventors or researchers improving a concept or engineers iteratively advancing a design. After a certain number of memetic evolution time loops, the memeplexes are forced to mix and new memeplexes are formed through a shuffling process. This shuffling enhances the quality of the memes after being infected by frogs from different regions of the swamp. Migration of frogs (cross-fertilization of ideas and/or designs) accelerates the searching procedure sharing their experience in the form of infection and it ensures that the cultural evolution towards any particular interest is free from regional bias. This major transfer of ideas can result in open academic forums or as new inventions and designs are made public.

Like all heuristics, parameter selection is critical to SFL algorithm performance. SFL algorithm has five parameters: the number m of memeplexes, the number n of frogs in a memeplex, the number q of frogs in a sub-memeplex, the number N of evolution or infection steps in a memeplex between two successive shufflings and the maximum step size allowed during an evolutionary step. Based on experience, the sample size (the number m of memeplexes multiplied by the number n of frogs in each memeplex), in general, is the most important parameter. An appropriate value for sample size is related to the complexity of the problem. The probability of locating the global (or near-global) optima increases with increasing sample size. However, as the sample size increases, the number of function evaluations to reach the goal increases, hence making it more computationally burdensome. In addition, when selecting m, it is important to make sure that n is not too small. If there are too few frogs in each memeplex, the advantage of the local memetic evolution strategy is lost. The response of the algorithm performance to q is that, when too few frogs are selected in a sub-memeplex, the information exchange is slow, resulting in longer solution times. On the other hand, when too many frogs are selected, the frogs are infected by unwanted ideas that trigger the censorship phenomenon that tends to lengthen the search period. The fourth parameter, N, can take any value greater than 1. If N is small, the memeplexes will be shuffled frequently, reducing idea exchange on the local scale. On the other hand, if N is large, each memeplex will be shrunk into a local optimum. The fifth parameter, the maximum step size allowed to be adopted by a frog after being infected. It actually serves as a constraint to control the SFL algorithm's global exploration ability. Setting this value small reduces the global exploration ability making the algorithm tend to be a local search. On the other hand, a large value may result in missing the actual optima, as it is not fine-tuned.

The general procedure of SFL algorithm is described below:

1. Generate randomly an initial population (P) of frogs (Solutions). For S-dimensional problems (S variables), a frog 'i' is represented as $X_i = (x_{i1}, x_{i2}, x_{iS})$.
2. Calculate the fitness value of each frog (solution).
3. Sort the frogs in a descending order of their fitness values.
4. Divide the entire population is into 'm' memeplexes, each containing 'n' frogs. Thus,
$$P = m \times n \tag{5.1}$$
5. Allocate the frogs to the memeflexes such that first frog goes to the first memeplex, the second frog goes to the second memeplex, likewise, frog 'm' goes to the mth memeplex, and frog '$m+1$' goes back to the first memeplex etc.
6. Within each memeplex, the frogs with the best fitness (X_b) and the worst fitness (X_w) are identified.
7. Apply the local search within memeplex using Eq. (5.2).
$$X_{i+1} = X_w + r \times (X_b - X_w) \tag{5.2}$$
r = random number between 0 and 1
If 'X_{i+1}' is better than 'X_i', replace current frog position 'X_i' with new position 'X_{i+1}'.
Else go to step 8.
8. Identify the frog with the global best fitness (X_g) and apply global search using Eq. (5.3)
$$X_{i+1} = X_w + r \times (X_g - X_w) \tag{5.3}$$
If 'X_{i+1}' is better than 'X_i', replace current frog position 'X_i' with new position 'X_{i+1}'.
Else go to step 9.
9. Generate new frog position 'X_i' randomly to replace the worst frog using Eq. (5.4).

$$X_{i+1} = X_{min} + r \times (X_{max} - X_{min}) \quad (5.4)$$

$X_{min} = (x_{min1}, x_{min2},.,x_{minS})$ and $X_{max} = (x_{max1}, x_{max2},., x_{maxS})$

10. Does the termination criteria met? If yes then go to step 11. Else go to step 2.
11. Stop.

    The flowchart for shuffled frog leaping algorithm is shown in Fig. 5.1.

5.2. Mechanism of Working of Shuffled Frog Leaping Algorithm

The mechanism of working of shuffled frog leaping algorithm is demonstrated through example $Min.z = x_1^2 - 3 \times 2 + 10$ such that $0 \leq x_1, x_2 \leq 3$.

Step 1: *Generate initial population*

The set of initial solutions and their fitness values for number of frogs (population size) = 12 is shown in Table 5.1.

Table 5.1: Initial population for SFL algorithm.

Frog no.	x_1	x_2	z
1	1.02	1.62	6.19
2	2.40	3.02	6.68
3	1.33	2.32	4.82
4	1.79	1.89	7.53
5	1.06	1.11	7.78
6	0.91	2.01	4.79
7	1.21	1.93	5.69
8	0.65	2.66	2.43
9	0.12	1.98	4.08
10	0.13	1.21	6.38
11	1.43	1.98	6.13
12	0.95	1.52	6.34

Step 2: *Sort the frogs in a descending order of their fitness values*

The frogs are sorted in a descending order of their fitness values as shown in Table 5.2. As shown in Table 5.2 frog 8 is at global best (X_g) position while frog 5 is at worst position.

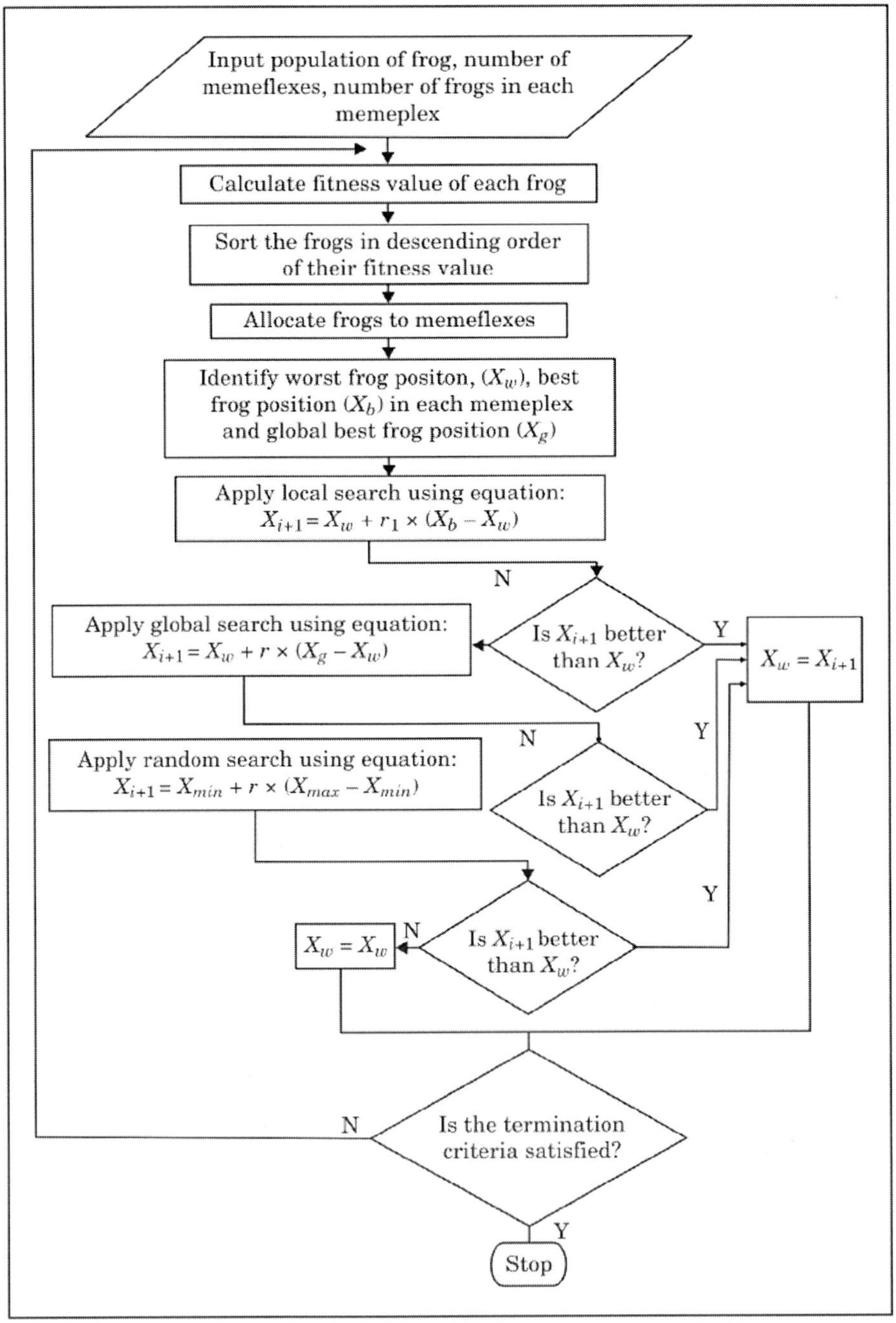

Fig. 5.1: Shuffled frog leaping algorithm (Eusuff *et al.,* 2006).

Table 5.2: Frogs sorted in a descending order of their fitness values.

Frog no.	x_1	x_2	z
8	0.65	2.66	2.43
9	0.12	1.98	4.08
6	0.91	2.01	4.79
3	1.33	2.32	4.82
7	1.21	1.93	5.69
11	1.43	1.98	6.13
1	1.02	1.62	6.19
12	0.95	1.52	6.34
10	0.13	1.21	6.38
2	2.40	3.02	6.68
4	1.79	1.89	7.53
5	1.06	1.11	7.78

Step 3: *Divide the entire population into memeplexes*

Let the frogs be divided into 3 memeplexes (m = 3) with each containing 4 frogs (n = 4). Allocate the frogs to the memeflexes such that first frog goes to the first memeplex, the second frog goes to the second memeplex, likewise, frog 'm' goes to the mth memeplex, and frog '$m+1$' goes back to the first memeplex etc. The allocation of frogs to the memeflexes is as shown in Table 5.3.

Table 5.3: Dividing frog into memeflexes.

	Memeflex 1				*Memeflex 2*				*Memeflex 3*		
Frog no.	x_1	x_2	z	*Frog no.*	x_1	x_2	z	*Frog no.*	x_1	x_2	z
8 (X_g & X_b)	0.65	2.66	2.43	9 (X_b)	0.12	1.98	4.08	6 (X_b)	0.91	2.01	4.79
3	1.33	2.32	4.82	7	1.21	1.93	5.69	11	1.43	1.98	6.13
1	1.02	1.62	6.19	12	0.95	1.52	6.34	10	0.13	1.21	6.38
2 (X_w)	2.40	3.02	6.68	4 (X_w)	1.79	1.89	7.53	5 (X_w)	1.06	1.11	7.78

As shown in Table 5.3, frog 8 is at global best position (X_g) and is also at best position (X_b) for memeflex 1, frog 2 is at worst position (X_w) in memeflex 1. Similarly, frog 9 and frog 4 are at best and worst positions respectively for memeflex 2, and frog 6 and frog 5 are at best and worst positions respectively for memeflex 3.

Step 4: ***Apply local search within memeflex***

Within each memeplex, the frogs with the best fitness (X_b) and the worst fitness (X_w) are identified. Now, apply the local search within memeplex using Eq. (5.2). If 'X_{i+1}' is better than 'X_i', replace current frog position 'X_i' with new position 'X_{i+1}'. After applying local search within each memeflex, the position of worst frog in each memeflex is updated as shown in Table 5.4.

Table 5.4: Updated positions of worst frog after applying local search

Memeflex 1			*Memeflex 2*			*Memeflex 3*		
x_{1new}	x_{2new}	Z_{new}	x_{1new}	x_{2new}	Z_{new}	x_{1new}	x_{2new}	Z_{new}
1.52	2.83	3.83	0.15	1.95	4.17	0.94	1.67	5.88

The updated positions of worst frog are then compared with their original positions and if the updated position is better than original position, the updated value replaces the original one. As shown in Table 5.4, since the updated values of worst frogs for all three memeflexes are better than their original values, the iteration is complete and new population is formed which is arranged in descending order of fitness values for the next generation as shown in Table 5.5.

Table 5.5: Updated population for second generation

Frog no.	*x1*	*x2*	*z*
8	0.65	2.66	2.43
2	1.52	2.83	3.83
9	0.12	1.98	4.08
4	0.15	1.95	4.17
6	0.91	2.01	4.79
3	1.33	2.32	4.82
7	1.21	1.93	5.69
5	0.94	1.67	5.88
11	1.43	1.98	6.13
1	1.02	1.62	6.19
12	0.95	1.52	6.34
10	0.13	1.21	6.38

Now, the new population is divided into memeflexes as shown in Table 5.6,

Table 5.6: Dividing frog into memeflexes

Memeflex 1				Memeflex 2				Memeflex 3			
Frog no.	x_1	x_2	z	Frog no.	x_1	x_2	z	Frog no.	x_1	x_2	z
8 (X_g& X_b)	0.65	2.66	2.43	2 (X_b)	1.52	2.83	3.83	9 (X_b)	0.12	1.98	4.08
4	0.15	1.95	4.17	6	0.91	2.01	4.79	3	1.33	2.32	4.82
7	1.21	1.93	5.69	5	0.94	1.67	5.88	11	1.43	1.98	6.13
1 (X_w)	1.02	1.62	6.19	12 (X_w)	0.95	1.52	6.34	10 (X_w)	0.13	1.21	6.38

When local search is applied, the updated positions of the worst frogs in three memeflexes are shown in Table 5.7.

Table 5.7: Updated positions of worst frogs after applying the local search.

Memeflex 1			Memeflex 2			Memeflex 3		
x_{1new}	x_{2new}	Z_{new}	x_{1new}	x_{2new}	Z_{new}	x_{1new}	x_{2new}	Z_{new}
0.88	2.36	3.70	1.29	1.59	6.91	0.12	1.39	5.85

It can be seen from Table 5.7 that, the worst frogs in memeflex 1 and memefles 3 are updated to better positions, however, the updated worst frog position for memeflex 2 is inferior than its original value, In such case, global search is applied using Eq. (5.3) and the result obtained is $x_{1new} = 0.78$, $x_{2new} = 1.95$ and $Z_{new} = 4.75$. Thus after applying global search, the updated value of worst frog is now superior to its original value. It is to be noted that, if the worst frog position does not improve even after global search, then the random search is applied as given by Eq. (5.4).

5.3. Optimization of Laser Beam Machining Process

Laser beam machining (LBM) is one of the most widely used thermal energy based non-contact type advance machining process. The mechanism of material removal during LBM includes different stages such as (i) melting (ii) vaporization and (iii) chemical degradation (chemical bonds are broken which causes the materials to degrade). When a high energy density laser bream is focused on work surface, the thermal energy is absorbed which heats and transforms the work volume into a molten, vaporized or chemically changed state that can easily be removed by flow of high pressure assist gas jet which accelerates the transformed material and ejects it from machining zone. The effectiveness of this process depends on thermal properties and, to a certain extent, the optical properties rather than the mechanical properties of the material to be machined. Therefore, materials that exhibit a high degree of brittleness, or hardness, and have favourable thermal properties, such as low thermal diffusivity and conductivity, are particularly well suited for laser machining. Since energy transfer between the laser and the material occurs through irradiation, no cutting forces are generated by the laser, leading to the absence of mechanically induced material damage, tool wear and machine vibration. Moreover, the material removal rate (MRR) for laser machining is not limited by constraints such as maximum tool force, built-up edge formation or tool chatter. LBM is a flexible process. When combined with a multi-axis work-piece positioning system or robot, the laser beam can be used for drilling, cutting, grooving, welding and heat treating processes on a single machine. The schematic of Nd:YAG Laser cutting system is as shown in Fig. 5.2.

In recent years, researchers have explored a number of ways to improve the LBM process performance by analyzing different factors that affect the quality characteristics of LBM process such as kerf width, kerf taper, surface finish etc. It is revealed from the literature that process performance of the LBM process can be improved considerably by appropriate selection of laser beam process parameters. Laser beam machining, being a non-contact process, does not involve any mechanical cutting forces and tool wear. Laser beam machining is a thermal energy based cutting process, which is executed by moving a focused laser beam on the surface of the work-piece with appropriate scanning speed. Asist gas is supplied through a nozzle with pressure to remove the molten metal. Laser beam machining of sheet metals has always been a major research area for getting the improves quality of cut which depend on several factors such as laser type, laser power, pressure of assist gas, cutting speed, sheet material composition and its thickness, and mode of operation of laser beam (Continuous or Pulsed mode). The capability of the laser cutting mainly depends on the optical and thermal properties rather than mechanical properties of the material to be cut. The materials which exhibit high degree of hardness or brittleness and passing through favorable thermal and optical properties such as low reflectivity, low thermal conductivity and diffusivity are well suited for laser cutting process. To achieve acceptable level of kerf quality characteristics, it is necessary to choose optimum combination of input process parameters.

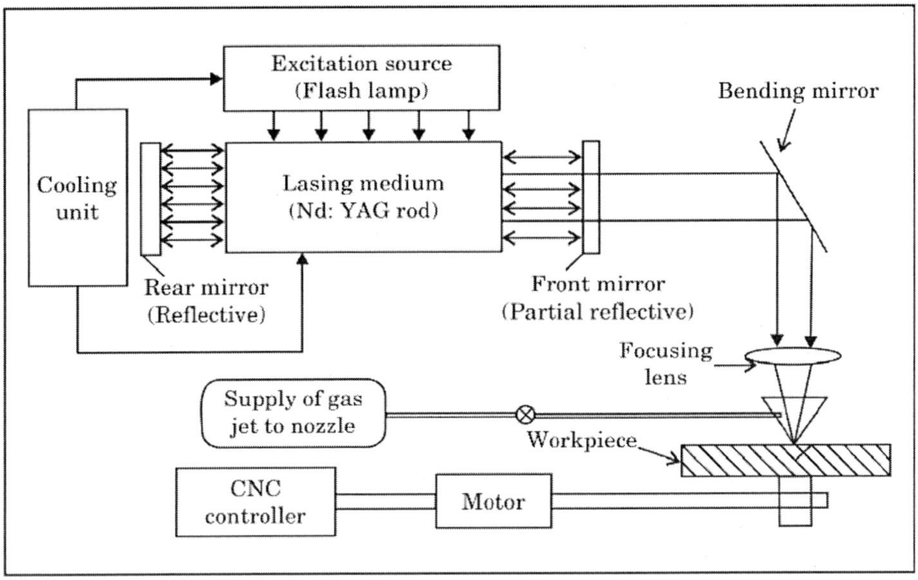

Fig. 5.2: Schematic of Nd: YAG Laser cutting system (Rao and Yadava, 2009).

***Kerf*:** The portion of material the laser burns away when it cuts through is called kerf.

Now, to demonstrate the effectiveness of SFL algorithm for optimization of laser beam machining process, an optimization model developed by Pandey and Dubey (2012) is discussed in this section. The objective is to minimize the kerf taper (*KT*) as given by Eq. (5.5).

Objective function:

$$Min: KT = 8.5670 - 2.528x_1 + 0.2093x_1^2 + 2.1318x_2^2 - 0.0371x_3^2 - 0.7193x_1x_2 + 0.0108x_3x_4 + 0.0752x_1x_3 \quad (5.5)$$

***Constraint*:** The constraint is on surface roughness value not to exceed 10.83 mm. surface roughness can be expressed by Eq. (5.6).

$$SR = -33.455 + 7.265x_1 + 12.191x_2 + 1.8114x_3 - 0.2813x_1^2 - 0.0726x_3^2 - 0.0055x_4^2 - 1.7719x_1x_2 \quad (5.6)$$

The combined objective function (z) considering both objective function and constraint function thus formulated as given by Eq. (5.7).

$$\text{Min.z} = \text{KT} + \text{penalty} \times (\text{Amount of constraint violation}) \quad (5.7)$$

Amount of constraint violation =

$$-33.455 + 7.265x_1 + 12.191x_2 + 1.8114x_3 - 0.2813x_1^2 - 0.0726x_3^2 - 0.0055x_4^2 - 1.7719x_1x_2 =$$

$$SR_i - 10.83 \text{ if } SR > 10.83$$

$$= 0, \text{ if } SR \leq 10.83 \quad (5.8)$$

The process variables and their bounds are given in Table 5.8.

Table 5.8: Variables and their bounds for optimization of laser beam machining process

Sl. no.	*Name of variable*	*Symbol*	*Unit*	*Lower bound*	*Upper bound*
1	Gas pressure	x_1	Kg/cm^2	5	9
2	Pulse width	x_2	ms	1.4	2.2
3	Pulse frequency	x_3	Hz	6	14
4	Cutting speed	x_4	Mm/min	15	25

The steps of SFL algorithm are applied to solve the above model as discussed below:

Step 1: *Generate initial population*

The set of initial solutions and their fitness values for number of frogs (population size) = 12 is shown in Table 5.9.

Table 5.9: Initial population

Frog no.	x_1	x_2	x_3	x_4	z
1	5.83	1.75	6.39	20.96	2.86
2	8.63	1.92	8.15	17.11	2.61
3	7.56	2.06	9.67	20.11	32.74
4	8.87	2.12	8.15	20.39	3.43
5	8.24	2.05	9.13	16.88	25.57
6	6.97	1.54	10.56	23.39	25.75
7	7.95	2.17	7.23	16.62	3.00
8	7.30	1.59	8.15	21.27	3.46
9	7.46	1.78	10.47	21.64	45.73
10	6.02	1.73	9.73	15.92	65.00
11	7.47	2.14	8.34	21.09	3.63
12	8.16	2.03	11.16	23.53	3.81

Step 2: *Sort the frogs in a descending order of their fitness values*

The frogs are sorted in a descending order of their fitness values as shown in Table 5.10. As shown in Table 5.10 frog 2 is at global best (X_g) position while frog 10 is at worst position.

Table 5.10: Frogs sorted in a descending order of their fitness values.

Frog no.	x_1	x_2	x_3	x_4	z
2	8.63	1.92	8.15	17.11	2.61
1	5.83	1.75	6.39	20.96	2.86
7	7.95	2.17	7.23	16.62	3.00
4	8.87	2.12	8.15	20.39	3.43
8	7.30	1.59	8.15	21.27	3.46
11	7.47	2.14	8.34	21.09	3.63
12	8.16	2.03	11.16	23.53	3.81
5	8.24	2.05	9.13	16.88	25.57
6	6.97	1.54	10.56	23.39	25.75
3	7.56	2.06	9.67	20.11	32.74
9	7.46	1.78	10.47	21.64	45.73
10	6.02	1.73	9.73	15.92	65.00

Step 3: *Divide the entire population into memeplexes*

Let the frogs be divided into 3 memeplexes (m = 3) with each containing 4 frogs (n = 4). Allocate the frogs to the memeflexes such that first frog goes to the first memeplex, the second frog goes to the second memeplex, likewise, frog 'm' goes to the m^{th}memeplex, and frog '$m+1$' goes back to the first memeplex etc. The allocation of frogs to the memeflexes is as shown in Table 5.11.

Table 5.11: Dividing frog into memeflexes

Memeflex 1					
Frog no.	x_1	x_2	x_3	x_4	z
2	8.63	1.92	8.15	17.11	2.61
4	8.87	2.12	8.15	20.39	3.43
12	8.16	2.03	11.16	23.53	3.81
3	7.56	2.06	9.67	20.11	32.74

Memeflex 2					
Frog no.	x_1	x_2	x_3	x_4	z
1	5.83	1.75	6.39	20.96	2.86
8	7.30	1.59	8.15	21.27	3.46
5	8.24	2.05	9.13	16.88	25.57
9	7.46	1.78	10.47	21.64	45.73

Memeflex 3					
Frog no.	x_1	x_2	x_3	x_4	z
7	7.95	2.17	7.23	16.62	3.00
11	7.47	2.14	8.34	21.09	3.63
6	6.97	1.54	10.56	23.39	25.75
10	6.02	1.73	9.73	15.92	65.00

As shown in Table 5.11, frog 2 is at global best position (X_g) and is also at best position (X_b) for memeflex 1, frog 3 is at worst position (X_w) in memeflex 1. Similarly, frog 1 and frog 9 are at best and worst positions respectively for memeflex 2, and frog 7 and frog 10 are at best and worst positions respectively for memeflex 3.

Step 4: *Apply local search within memeflex*

Within each memeplex, the frogs with the best fitness (X_b) and the worst fitness (X_w) are identified. After applying local search within each memeflex using Eq. (5.5), the updated position of worst frog in each memeflex is shown in Table 5.12.

Table 5.12: Updated positions of worst frog after applying local search

Local search applied to:	***Frog no.***	x_1	x_2	x_3	x_4	z
Memeflex 1	3	8.17	1.98	8.22	18.32	2.77
Memeflex 2	9	6.74	1.76	6.98	21.61	2.48
Memeflex 3	10	7.37	2.00	9.41	15.97	75.83

The updated positions of worst frog are compared with their original positions and if the updated position is better than original position the updated value replaces the original one. As shown in Table 5.12, the updated values of worst frogs for memeflexes 1 and 2 are better than their original values. However, the updated values of worst frogs for memeflexes 3 is inferior to that of its original value, global search is applied to worst frog in memeflex 3.

Step 5: ***Apply global search***

Since the local search does not improve the fitness value of worst frog in memeflex 3 (*i.e.,* frog 10), the global search is applied using Eq. (5.3). After applying global search, the updated position of frog 10 is $x_1 = 7.52$, $x_2 = 1.77$, $x_3 = 9.28$, $x_4 = 16.95$ with $z = 72.71$. It means that the solution 10 could not improve even after the global search. Hence, the random search is applied.

Step 6: ***Apply random search***

The random search is applied using Eq (5.4). After applying random search, the updated position of frog 10 is $x_1 = 5.68$, $x_2 = 1.51$, $x_3 = 13.97$, $x_4 = 16.62$ with $z = 38.11$. Now this updated value is better than original value (z=65.00). Values of the worst frog in all memeflexes are thus improved. The new population with the updated values of worst frogs arranged in the descending order of their fitness values is as shown in Table 5.13.

It may be noted that frog 9 was at worst position for memeflex 2 initially, after first generation is at global best (*gBest*) position. Also, the average fitness value of the population is significantly improved from 19.52 for initial solution to 9.79 after first generation. Now, the steps 3 to 6 are repeated till convergence.

Table 5.13: Updated population at the end of first generation.

Frog no.	x_1	x_2	x_3	x_4	z
9	6.74	1.76	6.98	21.61	2.48
2	8.63	1.92	8.15	17.11	2.61
3	8.17	1.98	8.22	18.32	2.77
1	5.83	1.75	6.39	20.96	2.86
7	7.95	2.17	7.23	16.62	3.00
4	8.87	2.12	8.15	20.39	3.43
8	7.30	1.59	8.15	21.27	3.46
11	7.47	2.14	8.34	21.09	3.63
12	8.16	2.03	11.16	23.53	3.81
5	8.24	2.05	9.13	16.88	25.57
6	6.97	1.54	10.56	23.39	25.75
10	5.68	1.51	13.97	16.62	38.11

5.4. Tool Path Planning for Hole Making Operations in Injection Moulds

Many industrial applications such as moulds, dies, engine block, automotive parts etc. requires machining of large number of holes. Hole making operations such as drilling, reaming, and tapping compose a large portion of machining processes for many industrial parts. For a part with many holes, a particular tool may be required by several holes and also tools of different diameters may be used to drill a single hole to its final size. Tool travel and tool switch planning are the two major issues in hole-making operations. To reduce tool traverse, it may be suggested that the spindle should not be moved until a hole is completely drilled using several tools of different diameters. This however will lead to excessive tool switches. By the same token, though tool switches can be reduced by completing all operations on all the holes that require the same tool, the travel time will be increased. Furthermore, the amount of tool movement and the number of tool switches will depend on which set of tools are to be used to drill each hole to its final size. The machining cost and tool cost are affected by the selection of tool combination for each hole. This gives rise to large number of possible sequences to complete hole-making operations on the part depending on location of hole and tool sequence to be followed. For example as shown in Fig.5.3, hole 3 may require one of {T_1, T_2, T_3}, {T_1, T_3}, {T_2, T_3}, and {T_3} tools to obtain the final size. Various combinations of tools for individual hole to achieve the desired size of hole has impact on optimum cutting speeds, tool switch time and tool travel distance.Hence, proper determination of the operations sequence and the corresponding machining speed used to perform each operation are crucial in reducing the total cost of production (Kolahan and Liang, 2000).

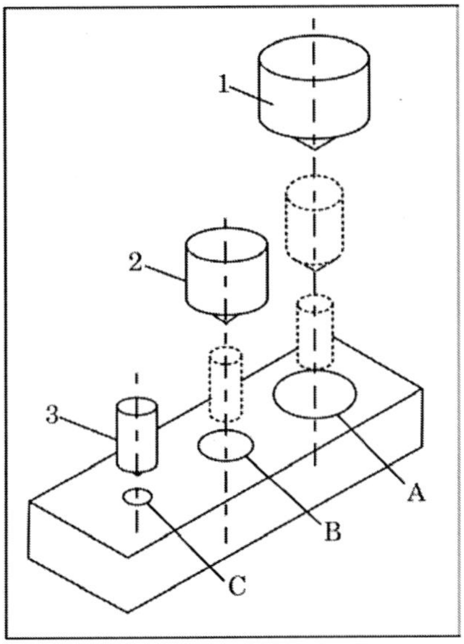

Fig. 5.3: Schematic representation of different set of tools for hole making (Kolahan and Liang, 2000).

In order to minimize the overall processing cost of hole-making operation, an optimization model is formulated based on analysis given by Kolahan and Liang (2000) and Ghaiebi and Solimanpur (2007) considering following components of total cost:

***Tool travel cost*:** Tool travel cost is the cost of moving the tool from its previous location to the current drilling position. Tool travel cost is proportional to the non-productive travelling distance required for the spindle to move between two drilling locations. It is assumed that, at a time, two axis drill press can travel only in one direction and a rectilinear distance function is used. The non-productive tool travel time a_{de} needed from hole d to hole e is given by:

$$a_{de} = |x_d - x_e| + |y_d - y_e| \tag{5.9}$$

Where (x_d, y_d) are co-ordinates of hole d and (x_e, y_e) are co-ordinates of hole e.

***Cost of Tool switching*:** It occurs whenever different tool is used for the next operation.

***Tool and machining costs*:** Tool cost includes the new tool cost and the cost of machine down time required to replace the tool. Machining cost comprises the operating cost and the machine overhead cost. Depth of cut, feed rate, and cutting speeds affect tool and machining costs. The actual combined tooling and machining costs when tool type d is used on hole e can be expressed as C_{de}.

Considering all components of cost, the optimization model is:

$$Min \sum_{d=1}^{D} \sum_{e=1}^{m_d} \sum_{\substack{d'=1 \\ d'\neq 1}}^{D} \sum_{e'=1}^{m_{d'}} \sum_{f=1}^{M-1} b * a_{de} x_{def} x_{d'e'f+1} +$$

$$\sum_{d=1}^{D} \sum_{e=1}^{m_d} \sum_{d'=1}^{D} \sum_{e'=1}^{m_{d'}} \sum_{j=1}^{M-1} a * S_{de,d'e'} \delta(T_{de}, T_{d'e'}) x_{def} x_{d'e'f+1} + C_{de} \tag{5.10}$$

Subject to

$$\sum_{f=1}^{M-1} x_{def} = 1, d = 1,2,\ldots..,D, e = 1,2,\ldots.m_i \tag{5.11}$$

$$\sum_{d=1}^{D} \sum_{e=1}^{m_d} x_{def} = 1, f = 1,2,\ldots.,M \tag{5.12}$$

$$x_{def} \leq \sum_{f'=f+1}^{M} x_{d,e+1,f'}, d = 1,2,\ldots..,D, e = 1,2,\ldots..m_i - 1 \tag{5.13}$$

$$x_{def} \in \{0,1), \forall d,e,f \tag{5.14}$$

Constraint given by Eq. (5.11) is to ensure that each operation of each hole is assigned to only one position in the sequence. Similarly, constraint given by Eq. (5.12) guarantee that only one operation is assigned to each position of the sequence.

Constraint given by Eq. (5.13) represent the precedence of operations of each hole. Constraints given by Eq. (5.14) confine the decision variables into zero-one values

The description of various terms used in this model is given below:

D is total number of holes to be machined in the part.

d, tool type index in ascending order according to the tool diameters, $d=1,...,D$

e, f, hole index, $e=1,...,E$ $f=1,...,E$

de, index for the last tool to be used on hole e

$C_{de,}$ combined tool and machining costs when tool type d is used on hole e.

a, cost per unit tool switch time

b, cost per unit non-productive travelling time

m_d the overall operations required for hole d., $d=1,2,\ldots,D$.

$M = m_1 + m_2 + \ldots + m_D$, the overall of operations in the part.

T_d the tool required for operation e of hole d.

$a_{dd'}$ the tool travel time for traveling from hole d to hole d'.

$S_{de,\, d'e'}$ the time required for switching the tool $T_{d'e'}$ when tool T_{de} is in spindle.

X_{def} = 1 if operation e of hole d is machined in position f of operation order, otherwise 0, where

$d=1,2,\ldots,D$, $e=1,2,\ldots, m_i$, $f=1,2,\ldots, M$.

$\delta(T_{de}, T_{d'e'}) = 1$ if T_{de}""$T_{d'e'}$, otherwise 0.

Now, to demonstrate the effectiveness of the SFL algorithm to solve above mathematical formulation, an industrial case study presented by Dalvi *et al.* (2016) is discussed below.

An application example of tool path planning in hole making operation is discussed in this section. An injection mould is considered in this example has 15 holes to be machined and finished as shown in Fig. 5.4. These holes are to machined using drilling, enlargement, and reaming operations. The identification numbers of different holes are shown in Fig. 5.5. The details of tools diameter are given in Table 5.14. Whenever a different tool is required for machining of a particular hole, the tool switch time is taken as 2 seconds. Table 5.15 presents combinations of tools required for machining a particular hole. Thus, for machining of Ø25 hole shown in Fig. 5.4, set of tools 1, 3, 4, 5 and 6 are to be used. Application example shown in Fig. 5.4 consists of machining of 15 holes through 47 hole-making operations. Thickness of mold plate is considered to be 80 mm.

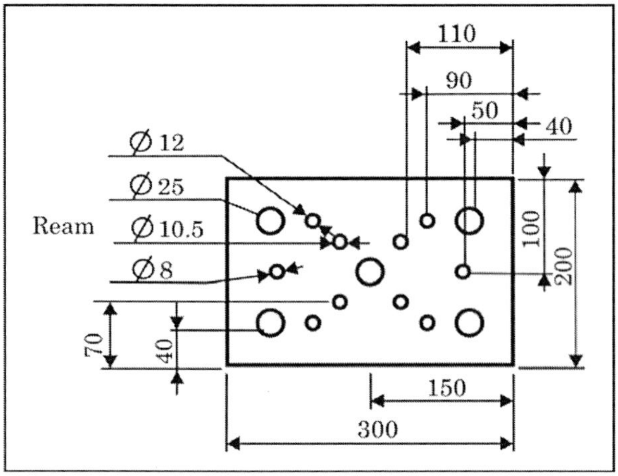

Fig. 5.4: Top view of example part

Table 5.14: Details of tools required hole-making for application example.

			Drill				*Reamer*
Tool type d	1	2	3	4	5	6	7
Tool Diameter (mm)	8	10	12	16	20	25	10.5

Table 5.15: Desired set of tools for machining individual hole in the application example.

Hole	Ø8	Ø10.5	Ø12	Ø25
Desired tools	1	1-2-7	1-3	1-3-4-5-6

Now, to solve this optimization model, following algorithm specific parameters for shuffled frog leaping algorithm are used:

- Frog population = 100
- Quantity of memeplexes = 10
- Quantity of sub frogs = 100
- Number of iterations = 100

Implementing the steps 1 to 11 of shuffled frog leaping algorithm discussed in section 5.1, the optimum sequence of hole making operations is obtained as given in Table 5.16.

Table 5.16: Optimal sequence hole-making operations using SFLA (Dalvi *et al.*, 2016)

Tool-Hole combination	8–H1	12–H1	16–H1	20–H1	25–H1	8–H2	12–H2	16–H2	20–H2
Tool-Hole combination	25–H2	8–H3	12–H3	8–H4	16–H3	20–H3	25–H3	12–H4	16–H4
Tool-Hole combination	20–H4	25–H4	8–H5	12–H5	16–H5	20–H5	25–H5	8–H6	12–H6
Tool-Hole combination	8–H8	12–H8	8–H9	8–H10	10–H10	10.5–H10	8–H7	8–H11	10–H11
Tool-Hole combination	8–H12	10–H12	10.5–H12	12–H9	8–H13	10–H13	8–H15	10.5–H13	10.5–H11
Tool-Hole combination	12–H7	8–H14							

Table 5.16 gives best possible sequence using shuffled frog leaping algorithm having optimum values of airtime of 2.083 minutes and tool switch time of 1.29 minutes. Considering a=Rs.25/min and b=Rs. 25/min, total processing cost of hole-making operations for the given application example is presented in Table 5.17.

Table 5.17: Total cost of hole making operations (Dalvi *et al.*, 2016).

Method	***C_{de} in Rs.***	***$b \times a_{de}$, in Rs.***	***$a \times S_{de}$, in Rs.***	***Total processing cost in Rs.***
SFLA	338	52.075	32.25	422.325

The SFL algorithm is thus a promising robust meta-heuristic to solve the problems in continuous as well as discrete domains. Also, like GAs and many other meta-heuristics, the SFL algorithm is very suitable for parallelization.

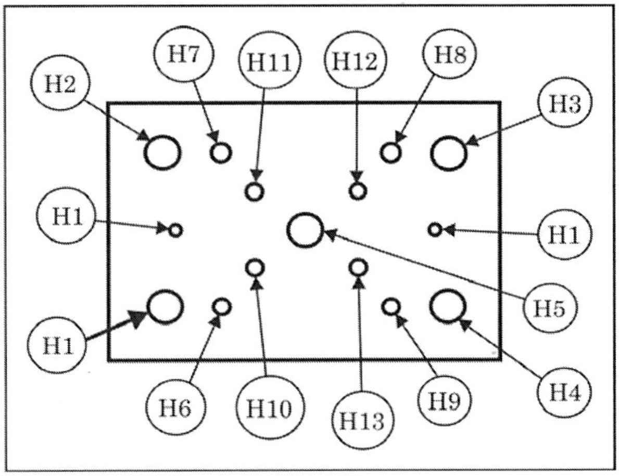

Fig. 5.5: Identification numbers for various holes.

5.5. Variants of Shuffled Frog Leaping Algorithm

Several variants of shuffled frog leaping algorithm are proposed by researchers to improve its performance as discussed below:

5.5.1. *Centroid mutated SFL algorithm*

SFL Algorithm despite having prominent features, it is sometimes criticized in terms of convergence rate and getting trapped in local optima for some computationally expensive functions. To enhance the convergence rate of basic SFL algorithm, Sharma *et al.* (2015) incorporated geometric centroid mutation (GCM) operator with a probability based mutation parameter named as C_m. GCM operator is stochastically applied, depending on the probability (C_m) in this approach. In each generation, if the randomly generated number between 0 and 1 is less than or equals to C_m then the new frog position is calculated according to the equation below otherwise the frog position is updated using Eq. (5.5). Mathematically geometric centroid mutation (GCM) expression using Eq. (5.15)

$$X_{i+1} = [(X_w + X_{r1} + X_{r2})/3] + r \times (X_b - X_w) \quad (5.15)$$

Where X_{r1}, X_{r2} are random frogs distinct and different from best and current frogs.

The mechanism Centroid Mutated shuffled frog leaping algorithm is explained below through demonstration example:

$$Min.\ z = x_1^2 - 3x_2 + 10 \text{ such that } 0 \le x_1, x_2 \le 3.$$

The initial solutions are considered same as those considered in section 5.2. The frogs sorted in descending order of their fitness values and population is divided in three memeflexes as shown in Table 5.18. Consider the probability $C_m = 0.45$ and let the randomly generated number is 0.32, then the Eq. (5.15) is applied to update the worst frog position in each memeflex.

Table 5.18: Population divided in three memeflexes

Memeflex 1				***Memeflex 2***				***Memeflex 3***			
Frog no.	x_1	x_2	z	***Frog no.***	x_1	x_2	z	***Frog no.***	x_1	x_2	z
8 (X_g& X_b)	0.65	2.66	2.43	9 (X_b)	0.12	1.98	4.08	6 (X_b)	0.91	2.01	4.79
3	1.33	2.32	4.82	7	1.21	1.93	5.69	11	1.43	1.98	6.13
1	1.02	1.62	6.19	12	0.95	1.52	6.34	10	0.13	1.21	6.38
2 (X_w)	2.40	3.02	6.68	4 (X_w)	1.79	1.89	7.53	5 (X_w)	1.06	1.11	7.78
X_{r1}	1.82	0.37			1.78	0.10			2.66	1.25	
X_{r2}	1.36	2.87			0.31	2.27			1.21	1.25	
Centroid	1.86	2.08			1.29	1.42			1.64	1.20	

Updated worst frog position is $x_{1new} = 1.51$, $x_{2new} = 3.10$ and $Z_{new} = 2.99$ for memeflex 1, $x_{1new} = 1.27$, $x_{2new} = 1.70$ and $Z_{new} = 6.50$ for memeflex 2, and $x_{1new} = 1.27$, $x_{2new} = 1.85$ and $Z_{new} = 6.05$ for memeflex 3. Since all updated positions are superior to original worst frog positions, the iteration is complete and new population arranged in descending order of fitness values for the next generation is as shown in Table 5.19.

It is observed that with centroid mutated approach, the average fitness value for the updated population at the end of first generation is 5.15 which is better than that of initial population (= 5.74) but inferior to that obtained using basic SFL algorithm (= 5.06). However, this mechanism could help the solution to jump out when getting trapped into local solution in subsequent generations.

Table 5.19: Updated population for second generation

Frog no.	x_1	x_2	z
8	0.65	2.66	2.43
2	1.51	3.10	2.99
9	0.12	1.98	4.08
6	0.91	2.01	4.79
3	1.33	2.32	4.82
5	1.31	2.09	5.46
7	1.21	1.93	5.69
11	1.43	1.98	6.13
1	1.02	1.62	6.19
12	0.95	1.52	6.34
10	0.13	1.21	6.38
4	1.27	1.70	6.50

5.5.2 Two phases learning SFL algorithm

In the local search of SFL Algorithm, the worst frog learns from the best frog in the same memeplex, the worst frog's possible position is substantially limited to the range between the current position and the best position, which restrains the search area of meme evolution, and decreases the convergence velocity; As the iteration increases, the worst frog become nearer to the best one in the same memeplex, it makes the algorithm to be premature, and the diversity becomes worse. In global information exchange, frogs of each memeplex are fixed and sorted in descending order, then algorithm repartitions memeplexex. However, the advantages of global optimal frog are not fully utilized, which influences the evolution effect. Based on the above reasons, Zhao and Lv (2015) proposed two-phases learning shuffled frog leaping algorithm using learning in the local search and in the global information exchange.

The best frog of each memeplex has better evolution advantage in local search. If the worst frog can learn from the best frog of other memeplex in the evolution process, it can bring three benefits. The first is to increase the possibility of the worst frog towards to the global optimal point; the second is to enlarge the search scope of the worst frog according frog rule; finally, it can ensure the diversity of each memeplex in the evolution process. The leaping step size of the worst frog is updated according to Eq. (5.16).

$$X_{i+1} = X_w + r_1 \times (X_{k,b} - X_{k,w}) + r_2 \times (X_{j,b} - X_{k,w}) \tag{5.16}$$

Where j and k are two different memeflexes. r_1 and r_2 are random numbers between 0 and 1. The worst from learns from best frog in its own memeflex is presented by the term $(X_{k,b} - X_{k,w})$ whereas, the worst frog learns from best frog in its other memeflex is presented by the term $(X_{j,b} - X_{k,w})$. Now, if the new position is better than previous position of worst frog, new position replaces the previous position, Otherwise the global search is applied using Eq. (5.17) below:

$$X_{i+1} = X_w + r_1 \times (X_g - X_{k,w}) + r_2 \times (X_{j,b} - X_{j,w}) \quad (5.17)$$

Where, X_g is global best position. If the solution still could not improve then random search is applied. The mechanism of Two phases learning SFL algorithm is explained below through our demonstration example:

$$Min.\ z = x_1^2 - 3x_2 + 10 \text{ such that } 0 \le x_1, x_2 \le 3.$$

Considering same initial solutions as those considered in section 2, the population is divided into three memeflexes as shown in Table 5.20.

Table 5.20: Population divided in three memeflexes

Memeflex 1				***Memeflex 2***				***Memeflex 3***			
Frog no.	x_1	x_2	z	***Frog no.***	x_1	x_2	z	***Frog no.***	x_1	x_2	z
8 X_g & X_b	0.65	2.66	2.43	9 (Xb)	0.12	1.98	4.08	6 (Xb)	0.91	2.01	4.79
3	1.33	2.32	4.82	7	1.21	1.93	5.69	11	1.43	1.98	6.13
1	1.02	1.62	6.19	12	0.95	1.52	6.34	10	0.13	1.21	6.38
2 (X_w)	2.40	3.02	6.68	4 (Xw)	1.79	1.89	7.53	5 (Xw)	1.06	1.11	7.78

Now applying local search for memeflex 1, k =memeflex 1 so $X_{k,b}$ = frog No.8 and $X_{k,w}$ = frog No.2. Let j= memeflex 3, so $X_{j,b}$ = frog No.6. Now using Eq. (5.16), the worst frog in memeflex 1 is updated to new position x_{1new} = 0.68, x_{2new} =2.74 and Z_{new} = 2.23. Similarly the updated position of worst frog in memeflex 2 is x_{1new} = 0.93, x_{2new} =2.35 and Z_{new} = 3.79 and that for memeflex 3 is, x_{1new} = 0.50, x_{2new} =1.79 and Z_{new} = 4.88. Since all updated positions are superior to original worst frog positions, the iteration is complete and new population arranged in descending order of fitness values for the next generation is as shown in Table 5.21.

Table 5.21: Updated population for second generation

Frog no.	x_1	x_2	z
2	0.68	2.74	2.23
8	0.65	2.66	2.43
4	0.92	2.35	3.79
9	0.12	1.98	4.08
6	0.91	2.01	4.79
3	1.33	2.32	4.82
5	0.50	1.79	4.88
7	1.21	1.93	5.69
11	1.43	1.98	6.13
1	1.02	1.62	6.19
12	0.95	1.52	6.34
10	0.13	1.21	6.38

It is observed that with two phase learning SFL algorithm, the average fitness value for the updated population at the end of first generation is 4.81 which is not only better than that of initial population (= 5.74) but also better than that obtained using basic SFL algorithm (= 5.06).

5.5.3. *Modified shuffled frog leaping algorithm*

In order to achieve better exploration and exploitation characteristics of standard SFL algorithm, Liang *et al.* (2016) modified standard SFL algorithm with respect to following aspects:

- The partition strategy in the standard SFLA is replaced by a random strategy, which improves the diversity and exploitation of frog population, and is benefited for global searching.
- The evolution of sub-memeplexes in SFL Algorithm is the process of worst frog to adjust its position, which is insufficient for the population evolution, especially for the frogs with better fitness. Therefore, the sub-memeplexes is cancelled so that all the frogs in memeplexes can take part in the evolution. Furthermore, the learning object in standard SFL Algorithm is the best frog in sub-memeplex or in swarm. This singularity may cause the

algorithm get into a local optimum. Therefore, in modified SFL Algorithm, the frogs learn from a randomly selected frog with a better performance to increase the local exploration ability.

- The leaping step in SFL algorithm aims to decrease the difference with the best position in the sub-memeplex or in the swarm. However, it tends to fall into the local optimum and the position states of the better frogs are difficult to be improved. Therefore, in order to enhance the global exploration, the velocity (representing the inertia motion) of the frog is introduced in the leaping step.

The steps of the modified shuffled frog leaping algorithm proposed by Liang *et al.* (2016) are as discussed below.

***Step 1*:** (swarm generation): Generate randomly an initial population (*P*) of frogs (Solutions). For S-dimensional problems (S variables), Calculate the fitness value of each frog (solution). Sort the frogs in a descending order of their fitness values.

***Step 2*:** (memeplexes partition): Generate a random numbers sequence, the length of which is the size of the frogs swarm; sort these random numbers, then get the order sequence; in turn select a number of frogs in this order sequence to form a memeplex.

***Step 3*:** (memeplexs evolution): According to aforementioned modification, the *i*-th frog's leaping step is modified as:

$$D_{i+1} = c_1 \times r_1 \times D_i + c_2 \times r_2 \times (X_r - X_w) \tag{5.18}$$

Where, D_i is the leaping step size in i^{th} generation and X_r is a randomly selected frog with better performance than X_w

$$X_{i+1} = X_w + D_{i+1} \tag{5.19}$$

In fact, the memeplexes evolution is a process of mutual learning among the members in group. Hence, the performance of the whole frog group is improved after such internal evolution. The difference between the best and worst frogs can be decreased.

***Step 4*:** (memeplexs shuffle): After the local search of each memeplex is finished, all memeplexs are shuffled in which the frogs are reorganized in descending order of fitness. Repeatedly divide the population into memeplexs and carry out local search process, until memetic evolution generation is reached.

Step 5: (lowliest place elimination): The worst frog with lowest fitness in the swarm will be eliminated, and a new frog is randomly generated to fill in the vocation. Repeat step 1 to step 4 until generations are reached

5.5.4. *Mnemonic shuffled frog leaping algorithm with co-operation and mutation*

Wang *et al.* (2015) proposed a Mnemonic Shuffled Frog Leaping Algorithm with Cooperation and Mutation, which is inspired by the competition and cooperation methods of different evolutionary computing, such as PSO and GA. This algorithm has two important features (i) Shuffled frog leaping algorithm with improved local search strategy and (ii) Cooperation and mutation to improve accuracy, which exhibits strong robustness and high accuracy for high-dimensional continuous function optimization. The first feature is inspired by PSO algorithm to improve the leaping rule by combining velocity updating equation of PSO. Second feature is inspired by GA to improve accuracy in the local exploration procedure through cooperation and mutation as in case of GA. The exploration process of Mnemonic Shuffled Frog Leaping Algorithm with Cooperation and Mutation is as shown in Fig. 5.6. These two features of the algorithm are discussed below:

i) New leaping rule formulated with improved local search strategy:

The new leaping rule is formulated as given by Eq. (5.20) below.

$$D_{i+1} = w \times D_i + c_1 \times r_1 \times (X_b - X_w) + c_2 \times r_2 \times (X_g - X_w) \quad (5.20)$$

The parameter w is the inertia weight that helps to regulate the search process. c_1 and c_2 are two random parameters in the range between 1 and 2. . The value of w is randomly selected, which makes history leaping step size impact on the current leaping step size of rendering randomness, leads to more flexibility for adjusting the global search and local search ability, and to a certain extent helps to maintain the diversity of the population.

$$w = 0.5 + [\text{rand}\ (0,\ 1)\ /2] \quad (5.21)$$

The worst frog position is then updated as:

$$X_{i+1} = X_w + D_{i+1} \quad (5.22)$$

(ii) Co-operation and mutation to improve accuracy

As shown in Fig. 5.6, cooperation between X_j and X_k is shown using three cross over points. The number of cooperation over points is also generated randomly. r_1, r_2 and r_3 are random numbers between 1 and number of variables. If the fitness value of new position offspring 1 (*off* 1) is better than that of X_j, offspring 1 will replace X_j. If the fitness value of new position offspring 2 (*off* 2) is better than that of X_k, offspring 2 will replace X_k.

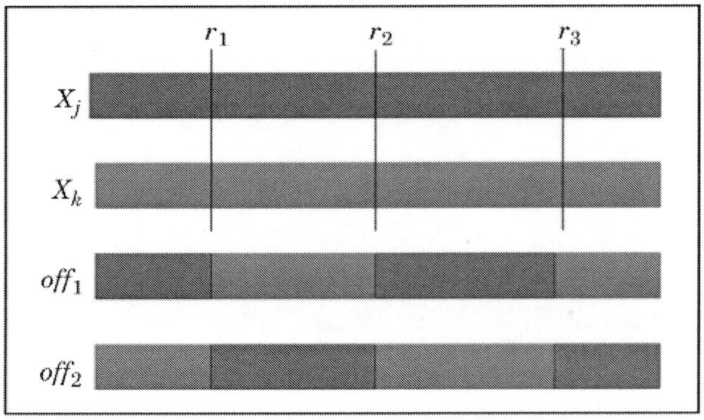

Fig. 5.6: Co-operation and mutation

To illustrate the process of co-operation, consider two randomly selected frogs (frog 4 of memeflex 1 and frog 8 of memeflex 2) from Table 5.11 of section 5.3. For frog 4, x_1= 8.87, x_2 = 2.12, x_3 = 8.15, x_4 = 20.39 with z = 3.43 and for frog 8, x_1= 7.30, x_2 = 1.59, x3 = 8.15, x_4= 21.27 with z = 3.46. Let the co-operation sites be 1 and 3. Then the offspring 1 will be x_1= 8.87, x_2 = 1.59, x_3 = 8.15, x_4 = 21.27 with z = 2.70. Since offspring 1 is superior to original frog 4, offspring 1 will replace frog 4. Similarly offspring 2 will be x_1= 7.30, x_2 = 2.12, x_3 = 8.15, x_4 = 20.39 with z = 3.52. Since offspring 2 is inferior to original frog 8, offspring 2 will not replace frog 8.

In the basic SFL Algorithm, the worst individual will be replaced by a random one when there is no improvement after learning from the local individual and the global best one. Although this method adds the diversity of the population, it ignores the beneficial information that exists in the population, so mutation is applied. It proposes a mutation method to which is $gbest_i$ mutation operation, and the variation equation is expressed as follows:

$$gbest_i = gbest_i + \xi N\ (0,\ 1)\ gbest_i \tag{5.23}$$

Where ξ is a variation of parameters; N (0, 1) is a random variable whose mean and variance are 0 and 1 respectively. $gbest_i$ is a global optimal solution after mutation. The above mutation method is applicable to the whole situation, and in order to make the produced new position after mutation applies to all the memeplexes, Cauchy mutation is proposed as given by Eq. (5.24) to update the worst frog position, if it is not improved by the above process.

$$X_w = X_b + \beta \times X_b \tag{5.24}$$

Where β is random number, which is generated by Cauchy mutation.

6

Harmony Search Algorithm

6.1. Introduction

Harmony Search (HS) was first developed by Geem *et al.* (2001). It was inspired by the observation that the aim of music is to search for a perfect state of harmony. This harmony in music is analogous to find the optimality in a process. On one hand, the perfectly pleasing harmony is determined by the audio aesthetic standard. A musician always intends to produce a piece of music with perfect harmony. On the other hand, an optimal solution to an optimization problem should be the best solution available to the problem under the given objectives and limited by constraints. Both processes intend to produce the best or optimum. Such similarities between two processes can be used to develop new algorithms by learning from each other. Harmony Search is just such a successful example by transforming the qualitative improvisation process into some quantitative rules by idealization, thus turning the beauty and harmony of music into an optimization procedure through search for a perfect harmony, namely, the Harmony Search (HS) or Harmony Search algorithm. The HS algorithm behaves excellent effectiveness and robustness when applied to several optimization problems and presents lots of advantages when compared to other heuristic optimization algorithms. The capability of the HS algorithm is sensitive awfully to parameter settings and new vector creating manner.

HS algorithm conceptualized a behavioral phenomenon of music players in improvisation process, where each player continues to polish its tune in order to produce better harmony. However, different from most emerging nature inspired algorithms, the inspiration of the HS is not from the natural phenomena, but is conceptualized from the musical process of searching for a perfect state of harmony determined by aesthetic standards. As we know, when musicians compose the harmony, they usually try various possible combinations of the music pitches stored in their memory. This kind of efficient search for a perfect harmony is analogous to the procedure of finding the optimal solutions to engineering problems. The HS method is inspired by the explicit principles of the harmony improvisation. Table 6.1 presents the comparison of harmony improvisation and optimization (Wang, 2009).

Table 6.1: Comparison of harmony improvisation and optimization (Wang, 2019).

Comparison factor	*Harmony improvisation*	*Optimization*
Targets	Aesthetic standards	Objective function
Best state	Fantastic harmony	Global optimum
Components	Pitches of the instrument	Values of the variables
Process units	Each practice	Each iteration

Apparently, the harmony memory consideration rate (HMCR) and pitch adjusting rate (PAR) are two basic parameters in the HS algorithm, which control the component of solutions and even affect convergence speed. The former is used to set the probability of utilizing the historic information stored in the harmony memory (HM). For example, 0.9 indicates that each component of a new solution will be chosen from the HM with 90 % probability, and 10 % probability from the entire feasible range. Each component of the solution is subject to whether it should be pitch-adjusted, which is determined by Pitch Adjusting Rate (PAR). 1-PAR means the rate of doing nothing. For example, a PAR of 0.3 indicates that the neighboring value will be chosen with 30 % probability. Similar to the genetic algorithm (GA), particle swarm optimization (PSO), and differential evolution (DE), the HS method is a random search technique. It does not require any prior domain information, such as the gradient of the objective functions. However, different from those population-based evolutionary approaches, it only utilizes a single search memory to evolve. Therefore, the HS method has the characteristics of algorithm simplicity. On the other hand, it occupies some inherent drawbacks, *e.g.,* weak local search ability. The optimization procedure using HS algorithm consists of steps 1–5, as follows:

Step 1: *Initialize the problem and algorithm parameters*

In this step, the optimization problem is specified in terms of objective functions, constraints, and decision variables along with their upper and lower bound values. The HS algorithm parameters are also specified in this step. These are the harmony memory size (HMS), or the number of solution vectors in the harmony memory; harmony memory considering rate (HMCR); pitch adjusting rate (PAR); number of decision variables (N) and the number of improvisations (NI), or stopping criterion. The harmony memory (HM) is a memory location where all the solution vectors (sets of decision variables) are stored. This HM is similar to the genetic pool in the GA. The HMCR and PAR are parameters that are used to improve the solution vector and are defined later in Step 3.

Step 2: *Initialize the harmony memory*

In Step 2, the HM matrix is filled with as many randomly generated solution vectors as the HMS:

$$HM = \begin{pmatrix} x_1^1 & x_2^1 & x_3^1 \ldots\ldots\ldots\ldots & x_N^1 \\ x_1^2 & x_2^2 & x_3^2 \ldots\ldots\ldots\ldots & x_N^2 \\ x_1^3 & x_2^3 & x_3^3 \ldots\ldots\ldots\ldots & x_N^3 \\ . & . & . \quad\quad . & . \\ . & . & . \quad\quad . & . \\ x_1^{HMS} & x_2^{HMS} & x_3^{HMS} \ldots\ldots\ldots\ldots & x_N^{HMS} \end{pmatrix} \quad (6.1$$

Step 3: *Improvise a new harmony*

A new harmony vector is generated based on three rules: (1) memory consideration, (2) pitch adjustment and (3) random selection. Generating a new harmony is called 'improvisation'. In memory consideration stage, the value of the first decision variable (x_1) for the new vector is chosen from any of the values in the specified HM range (x_1-x_{HMS}). Values of the other decision variables are chosen in the same manner. The HMCR, which varies between 0 and 1, is the rate of choosing one value from the historical values stored in the HM, while (1-HMCR) is the rate of randomly selecting one value from the possible range of values, as shown in the following equation:

$$\text{If HMCR} > \text{rand}\ (),\ x_i^{'} \in \left\{x_i^1, x_i^2, \ldots..x_i^{HMS}\right\}$$

$$\text{Else}\ x_i^{'} \in X_i \tag{6.2}$$

Where, rand () is a random number between 0 and 1. For example, a HMCR of 0.85 indicates that the HS algorithm will choose the decision variable value from historically stored values in the HM with an 85 % probability or from the entire possible range with a (100–85) % probability. Every component obtained by the memory consideration is examined to determine whether it should be pitch-adjusted. This operation uses the PAR parameter, which is the rate of pitch adjustment as follows:

$$\text{If PAR} > \text{rand}\ (),\ x_i^{'} = x_i^{'} \pm rand\ () \times bw \tag{6.3}$$

$$\text{Else}\ x_i^{'} = x_i^{'}$$

Where, *bw* is an arbitrary distance bandwidth. In this step, harmony memory consideration and pitch adjustment are applied to each variable of the new harmony vector one by one.

Step 4: ***Update harmony memory***

If the new harmony vector has better fitness function than the worst harmony in the HM, the new harmony is included in the HM and the existing worst harmony is excluded from the HM.

Step 5: Repeat Steps 3 and 4 until the termination criterion is satisfied

The flow chart of HS algorithm is shown in Fig. 6.2.

6.2. Mechanism of Working of Harmony Search (HS) Algorithm

The mechanism of harmony search algorithm is explained through demonstration example: $Min. z = x_1^2 - 3x_2 + 10$ such that $0 \leq x_{1,}x_2 \leq 3$.

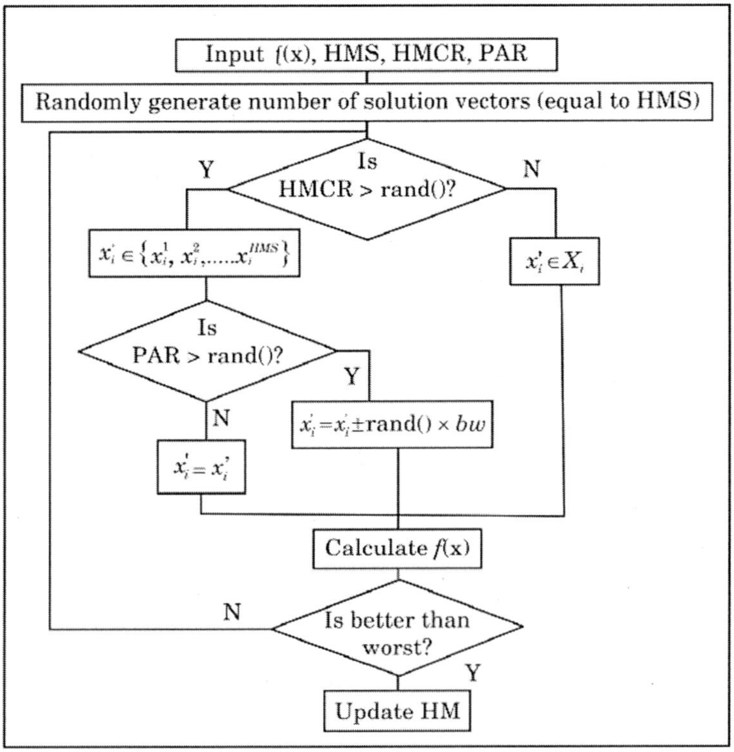

Fig. 6.2: Flowchart of HS algorithm (Geem *et al.*, 2001).

Step 1: ***Initialize the problem and algorithm parameters***

Let the algorithm specific parameters of HS algorithm set for this problem are:

- Harmony memory size (HMS): 10
- Harmony memory consideration rate (HMCR): 0.85
- Pitch adjusting rate (PAR): 0.35

Step 2: ***Initialize the harmony memory***

Table 6.2 shows harmony memory matrix.

Table 6.2: Harmony memory matrix.

Solution no.	x_1	x_2	z
1	1.59	0.27	11.69
2	2.43	1.35	11.84
3	**2.37**	1.86	10.04
4	1.33	2.32	4.82
5	0.90	0.05	10.64
6	2.71	**1.39**	13.15
7	2.38	1.42	11.41
8	0.33	0.19	9.54
9	0.65	2.66	2.43
10	0.12	1.98	4.08

Step 3: ***Improvise a new harmony***

To improvise New Harmony first rule of memory consideration is applied. Let the random number generated be 0.43. Since this number is less than HMCR, the value of the decision variables for the new vector is chosen from existing harmony memory as indicated in bold. The selected decision variables are x_1 = 2.37 and x_2 = 1.39 with z = 11.45. If the random number is greater than HMCR, a randomly generated solution, if superior, would replace the worst solution in harmony memory. The next rule is to apply pitch adjustment to the newly selected vector. Pitch adjustment is carried out by tuning the pitch within a given bandwidth. A small random amount is added to or subtracted from an existing pitch (or solution) stored in HM. Essentially, pitch adjustment is a refinement process of local solutions. Both memory consideration and pitch adjustment ensure that good local solutions are retained while the randomization makes the algorithm to explore global search space effectively. The subtlety is the fact that HS operates controlled diversification around good solutions, and intensification as well. The randomization explores the search space more widely and efficiently; while the pitch adjustment ensures that the newly generated solution is good enough, or not too far from existing good solutions. The intensification in the HS algorithm is represented by the harmony memory accepting rate. A high harmony acceptance rate means that good solutions from the history/memory are more likely to be selected or inherited. This is equivalent to a certain degree of elitism. Obviously, if the acceptance rate is too low, solutions will converge more slowly. As mentioned earlier, this intensification is enhanced by the controlled pitch adjustment. Such interactions between various components could be another important factor for the success of the HS algorithm over other algorithms. In addition, the structure of the HS algorithm is relatively easier. This advantage makes it very versatile to combine HS with other metaheuristic algorithms. For algorithm parameters, there are some evidences to suggest that HS is less sensitive to chosen parameters, which means that we may not have to fine-tune these parameters to get quality solutions. Furthermore, the HS algorithm is a population-based metaheuristic, which means that a group of multiple harmonies can be used in parallel. Proper parallelism usually leads to better performance with higher efficiency. The good combination of parallelism with elitism as well as a fine balance of intensification and diversification is the key to the success of the HS algorithm (Yang, 2008).

Now, let another random number generated is 0.31. Since it is less than PAR, the pitch adjustment is done using Eq. (6.4):

$$x_i' = x_i' \pm rand\ () \times bw \tag{6.4}$$

Considering Bandwidth b_w as $(x_{max}-x_{min})$ /6, the updated solution is: x_1 =2.19, x_2 =1.33 with z = 10.78. Being superior, this updated solution will replace the worst solution in initial population (x_1 =2.71, x_2 =1.39 with z = 13.15). If the random number is more than PAR, then selected harmony will not undergo pitch adjustment and if superior, will replace the worst harmony.

Step 4: *Update harmony memory*

The harmony memory will be now updated as:

Solution no.	x_1	x_2	z
1	1.59	0.27	11.69
2	2.43	1.35	11.84
3	2.37	1.86	10.04
4	1.33	2.32	4.82
5	0.90	0.05	10.64
6	2.19	1.33	10.78
7	2.38	1.42	11.41
8	0.33	0.19	9.54
9	0.65	2.66	2.43
10	0.12	1.98	4.08

Step 5: ***Repeat Steps 3 and 4 until the termination criterion is satisfied***

6.3. Motion Planning for Redundant Robot Manipulator under the Condition of Restrictions

The applications of industrial robots in the field of manufacturing are increasing due to their autonomy, flexibility, and self-governing work in different environments. In most of the manufacturing applications these robots has to perform tasks under the conditions of restrictions. However, under the condition of restrictions, with limited degrees of freedom, (*i.e.,* non-redundant robots) working zone is not achievable as shown in Fig. 6.3. To overcome this problem, manipulator with additional degrees of freedom called redundant manipulator is required as shown in Fig. 6.4. Motion planning of the redundant manipulator under condition of restrictions is typically an inverse kinematics problem. The main issue in solving inverse kinematic problems of a redundant manipulator lies in the fact that they are highly non-linear and the number of inverse kinematic solutions may become infinite and closed form solutions are impossible to find in general.

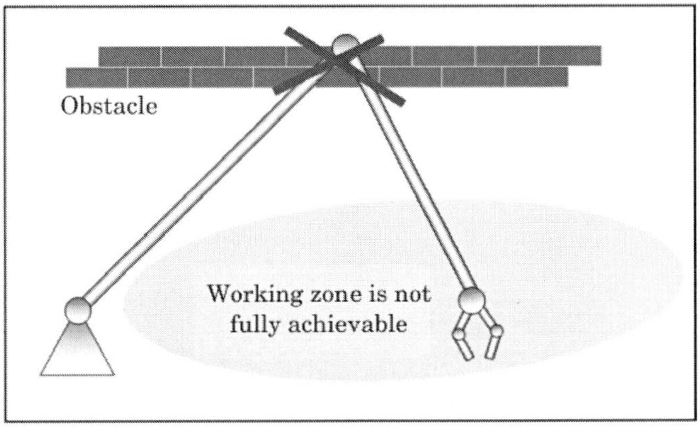

Fig. 6.3: Limited freedom degree manipulator with obstacle.

To solve this problem the objective is thus to minimize the positional error as the end effectors moves from one position to the other and the constraint is to ensure the collision free movement of the end effector. The collision free movement is obtained by systematically minimizing the overlap between the obstacle and the manipulator link. The positional accuracy is then obtained by using the HS algorithm.

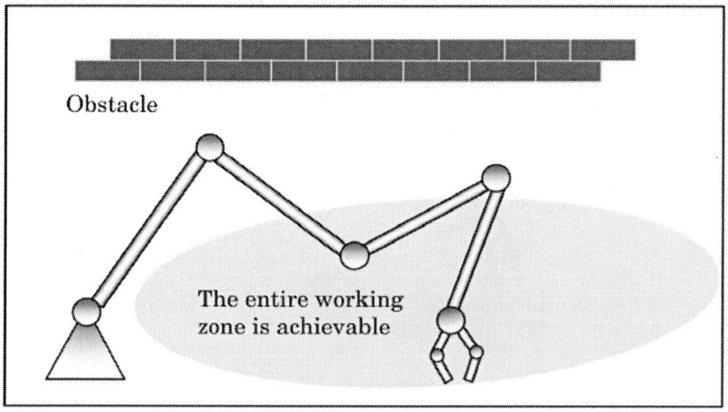

Fig. 6.4: Manipulator with additional degrees of freedom.

Evaluation of the objective function

The decision variables considered for this model are the angular positions of the links (θ_1, θ_2 ...θ_N), where N is the number of links of manipulator. The objectives function is to minimize the positional error (P_e) expressed by Eq. (6.5).

$$P_e = \sqrt{(x_c - x_f)^2 + (y_c - y_f)^2} \tag{6.5}$$

(x_c, y_c) is the current position of robot's end effector and (x_f, y_f) is the desired final position.

x_c and y_c can be expressed in terms of variables given by Eqs. (6.6) and (6.7) respectively.

$$x_c = \sum_{i=1}^{N} L_i \cos\theta_i \tag{6.6}$$

$$y_c = \sum_{i=1}^{N} L_i \sin\theta_i \tag{6.7}$$

Where, L_i and θ_i represents the length and orientation of the ith link respectively as shown in Fig. 6.5.

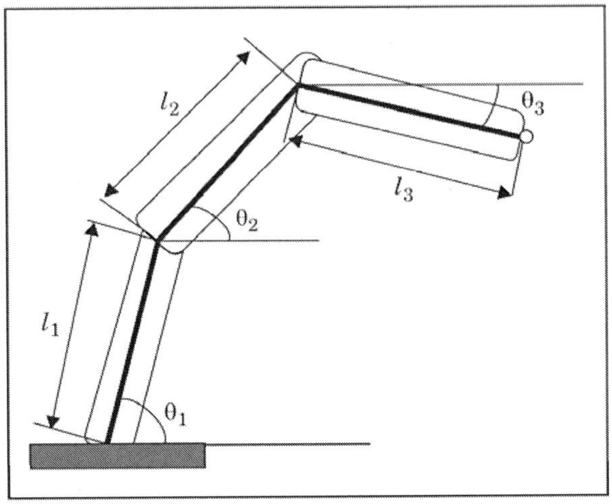

Fig. 6.5: Angular positions of the links in a manipulator.

Evaluation of constraint

The constraint is on collision free movement of manipulator. The collision is detected if overlap occurs between the manipulator link and the obstacle. The overlap area (A_o) can be calculated using following steps:

Step 1: Identification of point of intersection

Step 2: Identification of boundary line segment

Step 3: Identification of Overlap Region

Step 4: ***Calculating Area of Overlap Region***

The details of steps 1-4 are provided in section 2.4.

Step 5: ***Determination of penalty value:***

Penalty methods are a certain class of algorithms for solving constrained problems. A penalty method replaces a constrained optimization problem by a series of unconstrained optimization problems whose solution ideally converge to the solution of the original constrained problem.

The penalty function is a measure of violation of the constraints. The measure of violation is non-zero when the constraints are violated and is zero in the region where the constraints are not violated. The combined objective function considering the constraint function is thus formulated as given by Eq. (6.8).

$$\text{Minimize } Z = P_e + \text{Penalty} \times A_o. \tag{6.8}$$

Following procedure may be used to decide appropriate the penalty (Rao *et al.,* 2010).

(i) Calculate values of P_e and A_o at all boundary points.

(ii) Select the solutions which violate the constraint

(iii) Calculate the ratio $(r) = P_e / A_o$ for all selected solutions in step (ii).

(iv) Select the maximum value (r_{max}) among the ratio 'r' calculated in step (iii).

(v) Assign a penalty as a value slightly higher than r_{max}.

Consider that a 4 link robot manipulator has to perform the task of picking the object from position $(X_o, Y_o) = (0, 555)$ and placing it at position $(X_f, Y_f) = (330, 193)$ under the condition of restrictions offered by object shown in Fig. 6.6. The details of manipulator links are shown in Table 6.3.

Table 6.3: Manipulator link specifications.

Link number (i)	***Length of the link (L_i), mm***	***Initial angle (θ_i)***
1	110	90
2	150	90
3	105	90
4	190	90

The angular positions of the links L_1, L_2, L_3 and L_4 with respect to the horizontal plane are θ_1, θ_2, θ_3 and θ_4 respectively such that $0 \leq \theta_1, \theta_2, \theta_3, \theta_4 \leq 180°$.

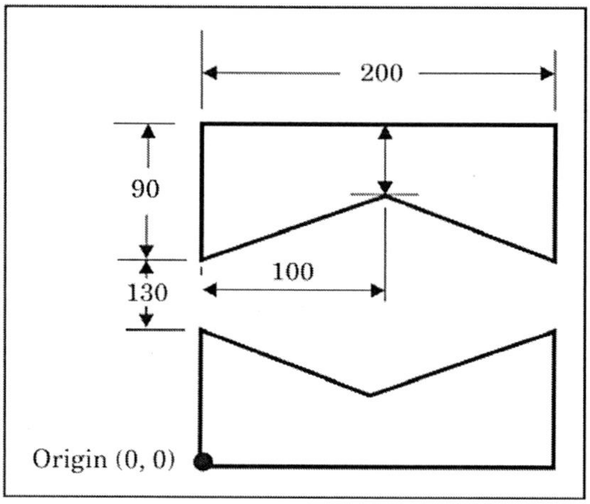

Fig. 6.6: Obstacle configuration (All dimensions are in mm).

Now, the steps of HS algorithm are applied to solve the above model:

Step 1: ***Initialize the problem and algorithm parameters***

Let the algorithm specific parameters of HS algorithm set for this problem are:

- Harmony memory size (HMS): 10
- Harmony memory consideration rate (HMCR): 0.85
- Pitch adjusting rate (PAR): 0.35

Step 2: ***Initialize the harmony memory***

The Table 6.4 shows harmony memory matrix.

Step 3: ***Improvise a new harmony***

To improvise New Harmony, first rule of memory consideration is applied. Let the random number generated be 0.67. Since this number is less than HMCR, the value of the decision variables for the new vector is chosen from existing harmony memory as marked bold in Table 6.4. The selected decision variables are: θ_1 = 70, θ_2 = 134, θ_3 = 142, and θ_4= 74. For this set of variables, using Eq. (6.8), the combined objective function value (z) is 502.55. Now, the solution is tested for pitch adjustment. Let the random number generated be 0.58. Since this random number is greater than PAR (=0.35) pitch adjustment will be not be done.

Table 6.4: Harmony memory matrix.

Sl. no.	θ_1	θ_2	θ_3	θ_4	x_c	y_c	P_e *(mm)*	A_o *(mm²)*	*Z*
1	**70**	89	69	67	152.45	526.17	377.53	0.00	377.53
2	175	101	11	134	–165.87	313.17	511.30	0.00	511.30
3	70	17	17	78	321.27	363.33	170.91	0.00	170.91
4	149	162	42	88	–151.13	361.34	511.22	0.00	511.22
5	3	88	**142**	5	220.33	234.56	116.77	22.54	1543.77
6	122	**134**	73	124	–236.46	460.56	626.98	0.00	626.98
7	17	40	90	**74**	272.74	414.71	230.36	59.33	3196.86
8	9	10	83	123	167.19	307.73	199.73	135.70	6984.73
9	64	110	156	168	–285.72	323.85	628.03	0.00	628.03
10	162	152	180	51	–222.11	253.29	555.47	0.00	555.47

The updated solution (θ_1 = 70, θ_2 = 134, θ_3 = 142, and θ_4 = 74 with z = 502.55) being better will replace the worst solution in initial population (θ_1 = 9, θ_2 = 10, θ_3 = 83, and q_4 = 123 with z = 6984.73). The improvised harmony memory matrix shown in Table 6.5.

Table 6.5: Improvise harmony memory matrix.

Sl. no.	θ_1	θ_2	θ_3	θ_4	x_c	y_c	P_e	A_o	*Z*
1	70	89	69	67	152.45	526.17	326.77	0.00	377.53
2	175	101	11	134	–165.87	313.17	413.93	0.00	510.22
3	**70**	17	17	78	321.27	363.33	175.93	0.00	170.55
4	149	162	42	88	–151.13	361.34	414.73	0.00	509.73
5	3	88	**142**	5	220.33	234.56	28.61	22.54	1244.28
6	122	**134**	73	124	–236.46	460.56	533.90	0.00	626.47
7	17	40	90	**74**	272.74	414.71	208.16	59.33	3195.48
8	70	134	142	74	–96.6	459	502.55	0.00	502.55
9	64	110	156	168	–285.72	323.85	533.02	0.00	629.47
10	162	152	180	51	–222.11	253.29	459.16	0.00	555.39

The initial position of manipulator and its position for solution 1 are shown in Fig. 6.7 (a) and Fig. 6.7(b) respectively.

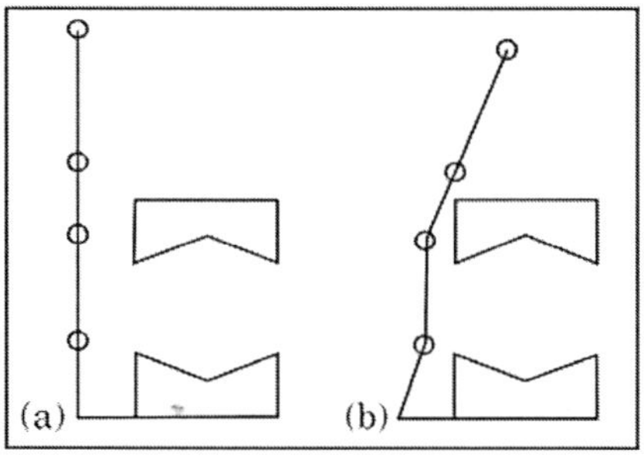

Fig. 6.7: Manipulator positions (*a*) Initial position and (*b*) position for solution 1

As shown in Table 6.5, solution 5 shows overlap with overlap area 22.54 mm^2.

Fig. 6.8 shows this overlap region.

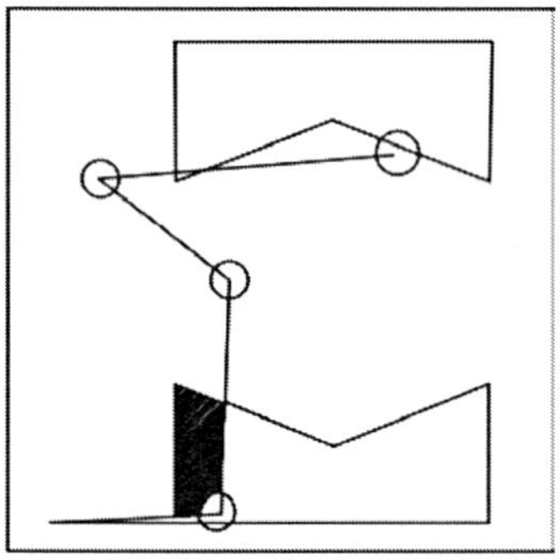

Fig. 6.8: Overlap region for solution 5.

Step 5: *Repeat Steps 3 and 4 until the termination criterion is satisfied*

The steps 3 and 4 are repeated till the termination criterion is satisfied at which the optimum solution is: $\theta_1 = 144°$, $\theta_2 = 29°$, $\theta_3 = 2°$, and $\theta_4 = 16°$ with $Z = 0.48$. The manipulator position for this optimum solution is shown in Fig. 6.9.

6.4. Parametric Optimization of Electro-Chemical Process

Electrochemical machining (ECM) is a metal-removal process based on the principle of metal removal by electro-chemical dissolution of the anode. Among various advanced machining processes, the electro-chemical machining process is one of the most highly developed processes. As shown in Fig. 6.10, the electro-chemical machining process uses a shaped tool (electrode). The tool is made the cathode and the work piece the anode. An electrolyte is pumped through the small gap which is maintained between the tool and the work piece. The chemical properties of the electrolyte are such that the constituents of the work material go into the solution by the electrolytic process but do not plate on the tool. By designing the tool accurately, the cavity or the hole which is to be produced exactly reproduces the tool shape.

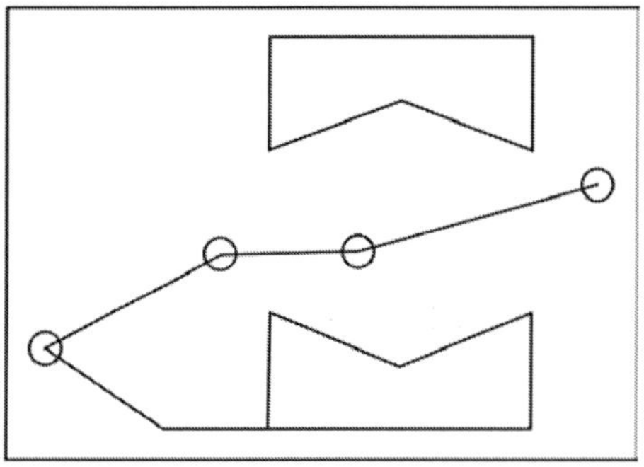

Fig. 6.9: Optimum manipulator position

The process can handle large variety of materials, limited only by their electro-chemical properties and not by their strength. This process is characterized by high metal removal rates for high strength and difficult-to-machine alloys. Fragile parts which otherwise not easily machinable can be shaped by electro-chemical machining process. Certain characteristics of electro-chemical machining process such as the ability to machine 3-dimensional curved surfaces without the striation marks, stress free and burr free machining, no thermal damage to the work piece, and ideally no tool wear, etc. make this process widely applicable.

The selection of optimum process parameters is very important issue in case of electro-chemical machining process due to following reasons:

- Electro-chemical process requires high initial investment along with high power consumption and large floor space requirement. Therefore use of this process is a costly affair.

- It is difficult to predict the changes that may occur in the inter-electrode gap (IEG). As the dimensional accuracy is controlled by IEG, optimum selection of parameters affecting IEG is very necessary.
- The parameters selected should control emission of heat and gas bubbles during process which otherwise may lead to variation in the electrolyte properties.
- Parameters selected should control pressure variation which otherwise may result into chocking of electrolyte flow.

This problem is further compounded by the corrosion, toxicity, and safety related problems of electro-chemical machining process.

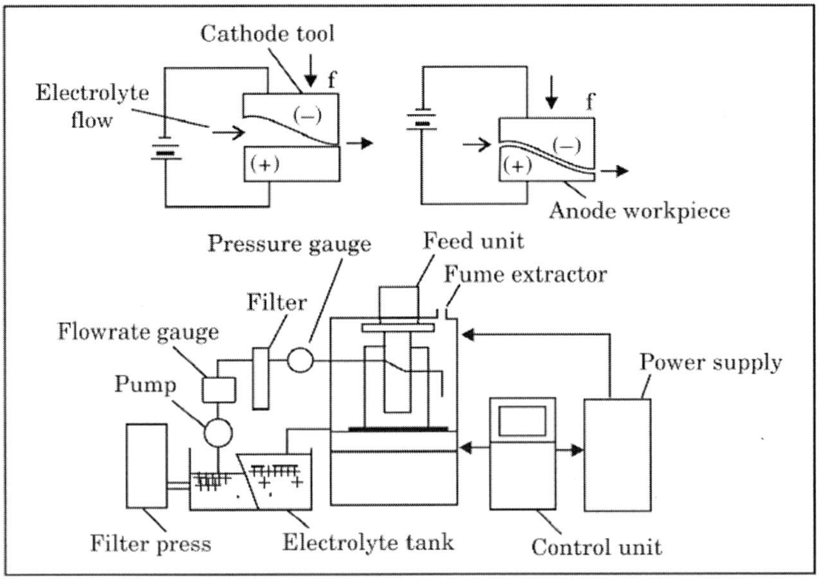

Fig. 6.10: Electro-chemical machining process (Rajurkar *et al.*, 1999).

Formulation of optimization model is the most important task in the optimization process. It involves identifying decision variables to be optimized, expressing the objective functions and constraints as functions of decision variables, setting up the bounds for decision variables, and finally expressing the optimization problem as a mathematical model in a standard format which can be directly solved by the optimization algorithm. The optimization model for electro-chemical machining process is formulated in the present work based on the analysis given by Acharya *et al.* (1986). The three decision variables considered for this model are, tool feed rate 'f' (μm/s), electrolyte flow velocity 'U' (cm/s), and applied voltage 'V' (volts).

This example present multi-objective optimization case considering three objectives namely, material removal rate, tool life, and dimensional inaccuracy subjected to the constraints of temperature, passivity, and chocking.

The first objective is to minimize the dimensional inaccuracy as given by following expression:

$$Z_1 = f^{0.381067} \times U^{-0.372623} \times V^{3.155414} \times e^{-3.128926} \tag{6.9}$$

Where, Z_1 = Dimensional inaccuracy (μm)

The second objective is to maximize the tool life by minimizing the number of sparks per millimetre, which is given by the following expression:

$$Z_1 = f^{3.528345} \times U^{0.000742} \times V^{-2.52255} \times e^{0.391436} \tag{6.10}$$

Where, Z_2 = number of sparks per millimetre

The third objective is to maximize the material removal rate, which is given by the following expression:

$$Z_3 = f \tag{6.11}$$

Where, Z_3 = Material removal rate (mm/s).

Constraints:

(a) Temperature constraint

To avoid boiling of electrolyte, the electrolyte temperature at the outlet should be less than the electrolyte boiling temperature. Mathematically this can be expressed as,

$$1 - f^{2.133007} \times U^{-1.088937} \times V^{-0.351436} \times e^{0.321968} \geq 0 \tag{6.12}$$

(b) Passivity constraint

Oxygen evolved during electro-chemical machining forms an oxide film, which is the root cause of passivity. To avoid passivity, the thickness of the oxygen gas bubble layer must be greater than the passive layer thickness. Mathematically this can be expressed as,

$$\left(f^{-0.844369} \times U^{-2.526076} \times V^{1.546257} \times e^{12.57697}\right) - 1 \geq 0 \qquad (6.13)$$

(c) *Choking constraint*

Hydrogen evolved at the cathode during ECM process can choke the electrolyte flow. To avoid chocking of the electrolyte flow, maximum thickness of the hydrogen bubble layer should be less than the equilibrium inter-electrode gap. Mathematically, it can be expressed as,

$$1 - f^{0.075213} \times U^{-2.488362} \times V^{0.240542} \times e^{11.75651} \geq 0 \qquad (6.14)$$

Parameter bounds for the variables are as follows:

$$8 \leq f \leq 200 \text{ (mm/s)};$$

$$300 \leq U \leq 5000 \text{ (cm/s)};$$

$$3 \leq V \leq 21 \text{ (Volts)}$$

The normalized combined objective function (Z) is formulated using different weights (w_1, w_2, and w_3) to three objectives (z_1, z_2, and z_3) as given by Eq. (6.15).

$$\text{Minimize } Z = \left(w_1 \times \frac{Z_1}{Z_{1_{min}}}\right) + \left(w_2 \times \frac{Z_2}{Z_{2_{min}}}\right) - \left(w_3 \times \frac{Z_3}{Z_{3_{max}}}\right) \qquad (6.15)$$

Where, $Z_{1_{min}}$ = minimum value of dimensional inaccuracy obtained when single objective optimization problem considering only dimensional inaccuracy as an objective, was solved for the given three constraints = 15.427 μm

$Z_{2_{min}}$ = minimum value of number of sparks per millimetre obtained when single objective optimization problem considering only tool life (in terms of number of sparks) as an objective was solved for the given three constraints = 1.055

$Z_{3_{max}}$ = maximum value of feed rate obtained when single objective optimization problem considering only material removal rate (in terms of feed rate) as an objective was solved for the given three constraints = 25 μm/s.

Now, the harmony search algorithm is applied to solve the above optimization problem. Considering equal weights to all objectives, the convergence graph obtained by applying steps 1-5 of HS algorithm for the above optimization model is shown in Fig. 6.11. The result of optimization is shown in Table 6.6.

Table 6.6: Result of multi-objective optimization.

Method	*Author*	*f* (*μl/s*)	*U* (*cm/s*)	*V* (*Volt*)	Z_1 (*μm*)	Z_2	Z_3 (*μm/s*)	*Z*
Goal programming	Acharya *et al.* (1986)	18.96	179	15	100	51.79	18.96	18.22
PSO	Rao *et al.* (2008)	8	300	9.835	39.34	3.39	8	1.811
HS		8	351	13.32	37.95	3.32	8	1.763

The results of multi-objective optimizations along with the previously published results using other methods are as given in Table 6.6 It is observed from the results that the combined objective function (*Z*) obtained by using HS method shows substantial improvement over Acharya *et al.* (1986) and Rao *et al.* (2008).

6.5. Variants of Harmony Search Algorithm

Several variants of harmony search algorithm are proposed by the researchers to improve its performance as discussed below:

6.5.1. *Modified HS algorithm*

As demonstrated in section 6.3 and section 6.4, although HS algorithm has been successfully applied to solve several real life problems, there are few drawbacks of the harmony search algorithm as given below:

- Harmony search algorithm requires large number of iterations to improvise upon the previous best solution.

- The selection of the appropriate bandwidth is a difficult task, due to which there is a chance that the decision variable may lie outside their boundary values and also of pre-mature convergence.

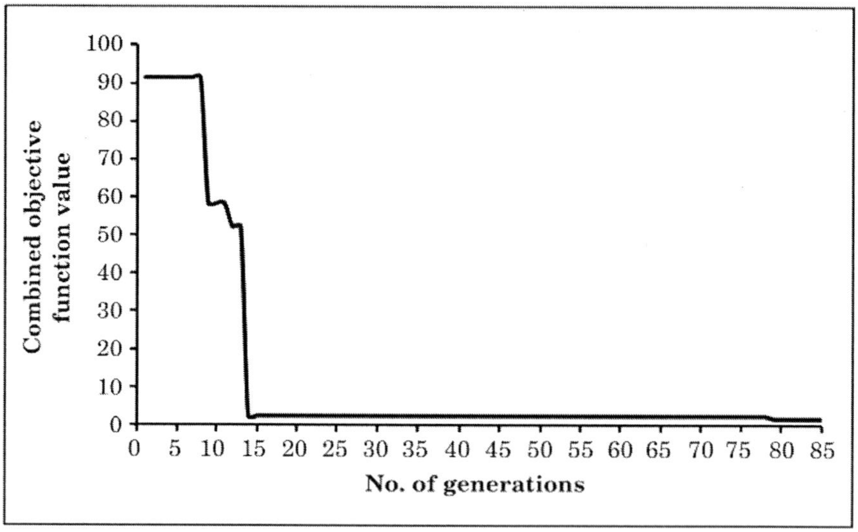

Fig. 6.11: Convergence graph for ECM process

To overcome the above drawbacks of the existing harmony search algorithm a modified harmony search algorithm is proposed by Pawar (2010). The mechanism of the proposed modified harmony search algorithm is discussed in next section.

In the proposed modified harmony search (HS_M) algorithm new harmony vector is based on four rules: (1) memory consideration, (2) previous best vector, (3) pitch adjustment, and (4) random selection. In conventional HS algorithm, the new vector is generated either by selecting randomly each decision variable from the specified HM range with probability HMCR or randomly selecting one value from the possible range of values with probability (1-HMCR). However, in conventional approach, although best solution is retained in harmony memory, it plays no direct role in generating new solution. Thus the algorithm creates a neighborhood solution of randomly generated vector from HM, which is normally a poor solution than the best sighted. Hence it requires large number of iterations to improvise upon the previous best solution. This process may also lead to local optimum solution. To overcome this limitation of conventional HS algorithm, in the proposed modified HS algorithm, the new vector is obtained as per the rules given in Eq. (2.9) below:

$$x_i^{'} \in \{x_i^1, x_i^2, ... x_i^{HMS}\} \text{ with probability } HMCR$$

$$x_i^{'} = x_{i-1}^{Best} \text{ with probability } (1\text{-}HMCR) \quad (6.16)$$

Where, x_{i-1}^{Best} is the best solution obtained in previous generation.

In conventional HS, if the pitch adjustment decision is yes, the decision variable is updated using Eq. (6.4). However, the selection of the appropriate bandwidth is a difficult task. Moreover there is a chance that the decision variable may lie outside their boundary values. To overcome this drawback of conventional HS algorithm, in the proposed algorithm if the pitch adjustment decision is yes, the decision variable is updated considering the standard deviation of the decision variable (s) as shown in Eq. (6.17).

$$x_i^{'} = x_i^{'} \pm \sigma \times (rand() - 0.5) \quad (6.17)$$

If the best previous solution is selected for updating (with probability (*1-HMCR*)), is also examined for pitch adjustment based on PAR. If the pitch adjustment decision is *yes*, each decision variable is updated as shown in Eq. (6.17). However, if the pitch adjustment decision is *no*, then the solution is updated by randomly selecting one value of each decision variable from the entire possible range of values of the corresponding variable. The modified harmony search algorithm is shown in Fig. 6.12.

The mechanism of modified HS algorithm is explained through demonstration example:

$$Min.\, z = x_1^2 - 3x_2 + 10 \text{ such that } 0 \le x_{1,}x_2 \le 3.$$

Steps 1 and 2 are same as those discussed in section 3.

Step 3: *Improvise a new harmony*

Let the random number generated be 0.94. Since this number is greater than HMCR, the best solution from the harmony memory which is solution number 9, having $x_1 = 0.65$, $x_2 = 2.66$ with z = 2.43 is selected. This solution must be pitch adjusted to find better solution in its vicinity. The pitch adjustment is carried out using Eq. (6.18).

$$x_i^{'} = x_i^{'} \pm \sigma \times (rand() - 0.5) \tag{6.18}$$

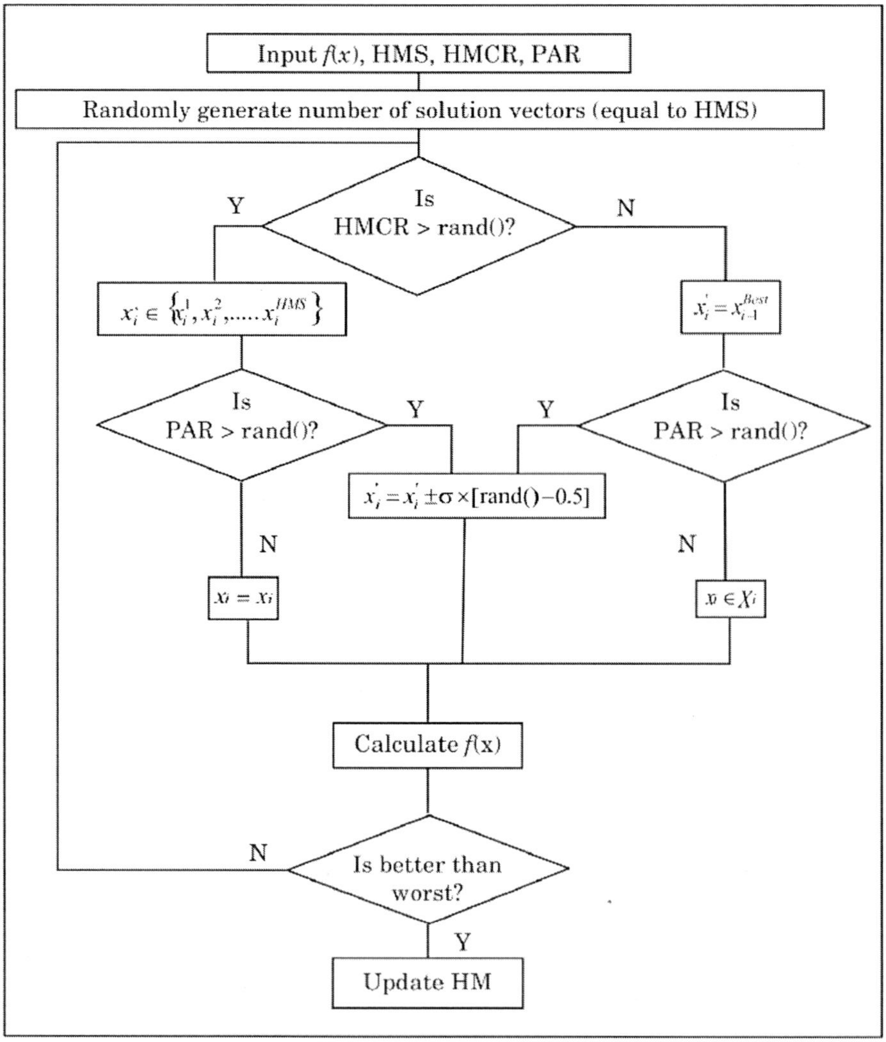

Fig. 6.12: Flowchart for modified harmony search algorithm (Pawar, 2010).

Where, σ = standard deviation = $\frac{(x_{max} - x_{min})}{6} = \frac{3}{6}$ = 0.5. The updated values of variables after pitch adjustment are x_1 = 0.46, x_2 = 2.78 with z = 1.86. Since better, this new solution will replace the worst solution in initial harmony memory. If this updated solution happens to be inferior to the worst solution in harmony memory then random solution is generated and tested for its superiority over the best solution. This strategy certainly help fast convergence at early stages of evolution without losing the beauty of original HS algorithm, as the probability of applying this strategy is only (1-HMCR). For example, in this particular case, it is only 15%.

Step 4: ***Update Harmony Memory***

The harmony memory will be now updated as shown in Table 6.7:

Table 6.7: Updated harmony memory

Solution no.	x_1	x_2	z
1	1.59	0.27	11.69
2	2.43	1.35	11.84
3	2.37	1.86	10.04
4	1.33	2.32	4.82
5	0.90	0.05	10.64
6	0.46	2.78	1.86
7	2.38	1.42	11.41
8	0.33	0.19	9.54
9	0.65	2.66	2.43
10	0.12	1.98	4.08

Steps 3 and 4 are repeated until the termination criterion is satisfied.

6.5.2. *Improved Harmony Search algorithm*

Mahdavi *et al.* (2007) proposed HS algorithm dynamic adaptation for both pitch adjustment rate (PAR) and bandwidth (*bw*) values. *PAR* and *bw* in HS algorithm are very important parameters in fine-tuning of optimized solution vectors, and can be potentially useful in adjusting convergence rate of algorithm to optimal solution. So fine adjustment of these parameters are of great interest. The traditional HS algorithm uses fixed value for both PAR and *bw*. In the HS method PAR and *bw* values adjusted in initialization step and cannot be changed during new generations. The main drawback of this method appears in the number of iterations the algorithm needs to find an optimal solution. Small PAR values with large *bw* values can cause to poor performance of the algorithm and considerable increase in iterations needed to find optimum solution. Although small *bw* values in final generations increase the fine-tuning of solution vectors, but in early generations *bw* must take a bigger value to enforce the algorithm to increase the diversity of solution vectors. Furthermore large PAR values with small *bw* values usually cause the improvement of best solutions in final generations which algorithm converged to optimal solution vector. The key difference between IHS and traditional HS method is in the way of adjusting PAR and *bw*. To improve the performance of the HS algorithm and eliminate the drawbacks due to fixed values of PAR and *bw*, IHS algorithm uses variables PAR and *bw* in improvisation step.The PAR value is linearly increased in each iteration of HS by using the following equation:

$$PAR_k = PAR_{min} + \frac{PAR_{max} - PAR_{min}}{NI} \times k \qquad (6.19)$$

Where, k= generation number, PAR_{min} and PAR_{max} are minimum and maximum values of pitch adjustment rates respectively and NI is total number of improvisations (k= 1, 2...NI).

The bandwidth (*bw*) value is exponentially decreased in each iteration of HS by using the following equation:

$$bw_k = bw_{max} \times e^{\frac{\ln\left(\frac{bw_{max}}{bw_{min}}\right)}{NI} \times k} \qquad (6.20)$$

Where, k= generation number, bw_{min} and bw_{max} are minimum and maximum values of bandwidths respectively and NI is total number of improvisations (k= 1, 2...NI).

6.5.3. *Harmony Search algorithm with differential mutation operator*

Chakraborty *et al.* (2009) proposed a new improvement to HS through inspiring the Differential Evolution (DE) mutation operator. Their proposed algorithm, named (DHS), replaces the pitch adjustment operation in classical HS with a mutation strategy borrowed from the DE algorithm. This step was accomplished through the mutation of the target vector with the difference of two randomly selected HM members (x_{r1} and x_{r2}) as shown in Eq. (6.21).

$$x_i^{'} = x_i^{'} + rand() \times (x_{r_1} - x_{r_2}) \tag{6.21}$$

It is worth mentioning here that the target vector is the new generated vector considering the memory selection step (with the probability of HMCR), and randomness (with the probability of (1-HMCR). Also the PAR operation is totally removed, thus, all new generated vector will be mutated according to Eq. (6.21).

6.5.4. *Adaptive Harmony Search Method*

Hasancebi *et al.* (2010) and Saka and Hasancebi (2009) proposed a new adaptation for HS by making both HMCR and PAR change dynamically during the improvisation process of HS. Goal of this dynamic adaptation is to encourage a more explorative search by allowing transitions to designs in the vicinity of the current solutions. This phenomenon is known as pitch adjustment in HS, and is controlled by PAR. In the standard algorithm both HMCR and PAR parameters are set to suitable constant values for all harmony vectors generated regardless of whether an exploitative or explorative search is indeed required at a time during the search process. On the contrary, in the adaptive algorithm a new set of values is sampled for HMCR and PAR parameters each time prior to improvisation of a new harmony vector, which in fact forms the basis for the algorithm to gain adaptation to varying features of the design space. Accordingly, to generate a new harmony vector in the algorithm proposed, a two-step procedure is followed consisting of: (*a*) sampling of control parameters and (*b*) Update harmony memory matrix and adaptivity.

(a) Sampling of control parameters

For each harmony vector to be generated during the search process, a new set of values are assigned for HMCR and PAR control parameters by probabilistically selecting them around average values of these parameters observed within the current harmony memory matrix using following equations.

$$(\text{HMCR})^K = \left(1 + \frac{(1 - HMCR)'}{(HMCR)'} . e^{-\gamma.N(0,1)}\right)^{-1} \quad (6.22)$$

$$(\text{PAR})^K = \left(1 + \frac{(1 - PAR)'}{(PAR)'} . e^{-\gamma.N(0,1)}\right)^{-1} \quad (6.23)$$

Where *(*HMCR*)*^{K} and *(*PAR*)*^{K} are the sampled values of the adapted parameters for a new harmony vector. *N(*0, 1*)* is a normally distributed random number, γ is the learning rate of adapted parameters, which is recommended to be selected within a range of [0.25, 0.50]. (HMCR)' and (PAR)' are the average values of improvisation parameters within the harmony memory matrix, obtained by averaging the corresponding values of all the solution vectors within the HM matrix, that are:

$$(HMCR)' = \frac{\sum_{i=1}^{HMS} HMCR^i}{HMS} \quad (6.24)$$

$$(PAR)' = \frac{\sum_{i=1}^{HMS} PAR^i}{HMS} \quad (6.25)$$

In general, these new set of values for (HMCR)^K and (PAR)^K are calculated based on probabilistic selecting from around average values of these parameters observed within the current harmony memory matrix.

In the proposed implementation, for each new vector a probabilistic sampling of control parameters is motivated around average values of these parameters (HMCR)' and (PAR)' observed in the harmony memory matrix. Considering the fact that the harmony memory matrix at an instant incorporates the best solutions (=HMS) sampled thus far during the search, the idea here is to encourage forthcoming vectors to be sampled with values that the search process has taken the most advantage in the past. The use of a logistic distribution provides an ideal platform in this sense due to two reasons. First, this distribution guarantees the sampled values of control parameters to lie within their possible range of variation, *i.e.,* (HMCR)^K, (PAR)^K. Second, it has a bell-shaped probability density function which is symmetrical around (HMCR)' or (PAR)'. It follows that values above and below (HMCR)' and (PAR)' are distributed with the same probability. This probability decreases significantly as it is moved away from (HMCR)' and (PAR)', indicating that small variations around the average value occur more frequently than larger ones.

(b) Update harmony memory matrix and adaptivity

Once the harmony memory matrix is updated, the HMCR and PAR parameters are recalculated. This way the harmony memory matrix is updated with the most recent information required for an efficient search and the forthcoming solution vectors are guided to make their own selection of control parameters mostly around these updated values. It is to be noted that there are no single values of control parameters that lead to the most efficient search of the algorithm throughout the process unless the design domain is completely uniform. On the contrary, the optimum values of control parameters have a tendency to change over time, depending on various regions of the design space in which the search is carried out. The update of the control parameters within the harmony memory matrix enables the algorithm to catch up with the varying needs of the search process as well. Hence, the most advantages values of control parameters are adapted in the course of time automatically *i.e.,* by the algorithm itself, which plays the major role in the success of adaptive HS method.

7

Simulated Annealing Algorithm

7.1. Introduction

The simulated annealing algorithm developed by Kirkpatrick *et al.* (1983) resembles the cooling process of molten metals through annealing. At high temperatures, the atoms in the molten metal can move freely with respect to each another, but as the temperature is reduced, the movement of the atoms gets restricted. The atoms start to get ordered and finally form crystals having the minimum possible energy. However the formation of crystal mostly depends on the cooling rates. If the temperature is reduced at very fast rates, the crystalline state may not be achieved at all; instead, the system may end up in polycrystalline states (by analogy, local optimum solution), which may have a higher energy state than the crystalline state (by analogy, global optimum solution), Therefore, in order to achieve the absolute minimum energy state, the temperature needs to be reduced at a slower rate.

The simulated annealing algorithm simulates this process of slow cooling of molten metal to achieve the minimum function value in the minimization problem. The cooling phenomenon is simulated by controlling a temperature like parameter introduced with the concept of Boltzman probability distribution. According to Boltzman probability distribution, a system in a thermal equilibrium at a temperature '*T*' has its energy distributed probabilistically according to Eq. (7.1).

$$P(E) = exp\ (-E\ /\ KT\ ) \qquad (7.1)$$

where '*K*' is Boltzman constant. This expression suggests that, a system at high temperature has almost uniform probability of being at any energy state. Therefore, by controlling the temperature '*T*' and assuming that the search process follows Boltzman probability distribution, the convergence of an algorithm can be controlled. At any current point x (t), the new value of the variables for the successive iterations is calculated using Eq. (7.2).

$$x(t+1) = x(t) + \sigma \times \left[\sum_{i=1}^{N} R_i - \frac{N}{2} \right] \tag{7.2}$$

Where, $\sigma = (x_{max} - x_{min}) / 6$, R = Random number and N = Number of random numbers used.

Using the Metropolis algorithm Metropolis *et al.* (1953), with current configuration '*i*', the probability of accepting '*j*' as a next configuration depends on the difference in the objective function values at these two points (*Δobj=obj (j)-Obj (i))* and is calculated using the Boltzman probability distribution given by Eq. (7.3):

$$Pr\{new = j | current = i\} = \begin{cases} 1 \; if \; \Delta obj \le 0 & \text{.......... Rule 1} \\ e^{-\frac{\Delta obj}{T}} \quad Otherwise & \text{.......... Rule 2} \end{cases} \tag{7.3}$$

If $\Delta obj \le 0$, this probability is one and the new configuration is always accepted. In the function minimization context, this makes sense because if the functional value at new configuration is better than the old one, the point new configuration must be accepted. The interesting situation happens when Δobj is bigger than zero, which implies that the function value at new configuration is inferior to that of previous configuration. According to many traditional algorithms, the point should not be chosen. However, according to the Metropolis algorithm, there is some finite probability of selecting the new configuration even though it is inferior to that of previous one. It is to be noted that this probability is not same in all situations. This probability depends on the magnitude of ΔT and T values as indicated by rule 2 of Eq. (7.3).

If the parameter T is large, this probability is more or less high for points with largely disparate functional values. Thus, any point is almost acceptable for a large value of T, on the other hand, if the parameter

T is small, then the probability of accepting an arbitrary point is small. Thus, for small values of *T*, the points with only small deviation in the function values are accepted.

Simulated annealing is a point-by-point method. The algorithm begins with an initial point and a high temperature *T*. A second point is created at random in the vicinity of the initial point and the difference in the function values [*Δobj*] at these two points is calculated. If the second point has a smaller function value, the point is accepted, otherwise the point is accepted with the probability of exp [-*Δobj*/*T*]. This completes an iteration of this simulated annealing procedure. In the next generation, another point is created at random in the neighborhood of the current point and the Metropolis algorithm is used to accept or reject the point. In order to simulate the thermal equilibrium, at every temperature the number of points (*n*) are usually tested at a particular temperature, before reducing the temperature. The algorithm is terminated when a sufficiently small temperature is obtained or a small enough change in function value is found. An estimate of the initial temperature can be obtained by calculating the average of the function values at a number of random points in the search space. Decrement factor is left to the choice of the user. Some of the features of the simulated annealing algorithm are as follows:

a) Because of the discrete nature of the function and construction evaluations, the convergence or transition characteristics are not affected by the continuity of differentiability of the function.
b) The convergence is also not influenced by the convexity status of the feasible space.
c) The design variables need not to be positive.
d) The method can be used to solve mixed-integer, discrete, or continuous problems.

The flowchart of SA algorithm is shown in Fig. 7.1.

7.2. Mechanism of Working of Simulated Annealing (SA) Algorithm

The mechanism of harmony search algorithm is explained through demonstration example: $Min. z = x_1^2 - 3x_2 + 10$ such that $0 \leq x_{1,}x_2 \leq 3$.

Step 1: ***Set initial conditions***

For simulated annealing algorithm, the algorithm specific parameters to set are: initial temperature (T_o), initial set of variable (X), decrement in temperature or cooling schedule (C_T), and termination temperature (T_e). The initial temperature has significant influence on the performance of simulated annealing algorithm. Usually, the initial temperature is considered as the average function value at all boundary points. As this example has two variables, number of boundary points will be four. Those are: (1) $x_1 = 0$, $x_2 = 0$ with $z = 10$; (2) $x_1 = 3$, $x_2 = 0$ with $z = 19$; (3) $x_1 = 0$, $x_2 = 3$ with $z = 1$; and (4) $x_1 = 3$, $x_2 = 3$ with $z = 10$. The average value of z at these four point is 10. Hence the initial temperature is considered as 10°C. The termination temperature is considered as 0.01°C. Let the set of initial solution be $x_1 = 1.80$ and $x_2 = 1.02$ with $z = 10.17$. The cooling schedule is taken as 0.1. It means for every successful move, the temperature is reduced by 10% of current state temperature.

Step 2: ***Generate new solution***

Considering any five random numbers (N=5) and the standard deviation for both variables as 0.5, using Eq. (7.2) the new solution generated in the vicinity of initial solution is: $x_1' = 1.65$ and $x_2' = 0.60$ with $z' = 10.91$. It is thus observed that the new solution is slightly inferior to initial solution. This new solution is then tested for feasibility (values of variables should be within their respective bounds). If the solution is feasible it is considered for further process, otherwise rejected and another solution is generated with new set of random numbers. In this case since the solution is feasible, it goes to the next step.

Step 3: ***Apply rules of acceptance of new solution***

The probability of accepting 'j' as next configuration depends on the difference in the function value at these two points (Δobj). Here, $obj = z' - z = (10.91–10.17) = 0.74$. Now, since the objective is to minimize and the simulated annealing algorithm is naturally suitable for solving minimization problem, the fitness values are same as those of function values. The probability of accepting new solution is then calculated using Eq. (7.3). Since $\Delta obj > 0$, second rule will apply. The probability of accepting new solution (P_r) using this rule is $e^{-\Delta obj/T} = 0.93$. Now let the random number generated (R) is 0.73. Since $P_r > R$, the new solution will be accepted. Thus, an important characteristics of SA algorithm is that, it gives chance to inferior solution also, with some probability, to evolve. This is not the case with most of the metaheuristics. However, the probability of acceptance will decreases

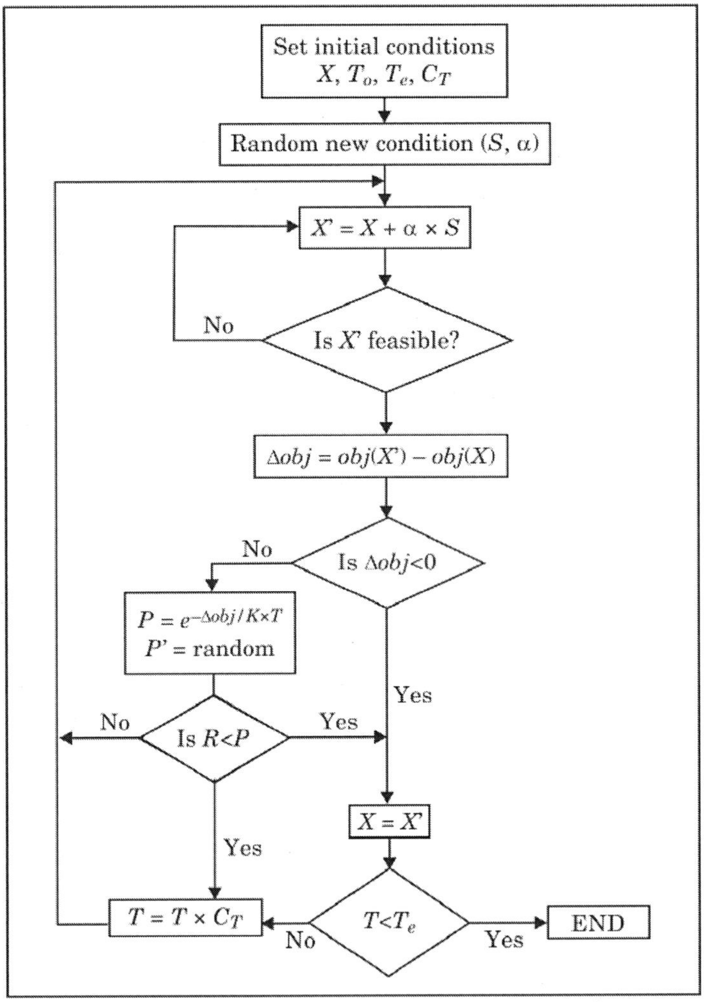

Fig. 7.1: Flowchart of simulated annealing algorithm (Kirkpatrick *et al.,* 1983).

as it evolve and after many successful moves when temperature is reduced substantially. For example, consider the same amount of inferiority (Δobj = 0.74) but at reduced temperature (T) = 2°C. The probability of accepting new solution (P_r) will now be 0.69 only and thus by rule 2, the new solution will be rejected since $P_r < P$. Thus the inferior solutions are allowed to enter into the next generation only during initial phase of evolution when the temperatures are high, this is thus in accordance with the annealing process in which the atoms makes random movement at high temperature but as the temperature reduces, the atoms get more structured and their movement is restricted. Now, if the new solution is accepted then the move is successful and the temperature is reduced by its decrement rate. Thus for this example, after the first generation, the temperature will be 9°C.

Step 4: ***Check whether termination criteria is met***

For simulated annealing, the termination criteria is 'End Temperature (T_e)'. The temperature obtained after first generation is 9°C which is greater than T_e (=0.01°C in this case). Hence the steps 2 to 4 are repeated until the temperature goes below T_e.

7.3. Design Optimization of a Universal Motor Using Simulated Annealing Algorithm

Universal electric motors are named after their ability to function both with alternating current and direct current. The most significant feature of universal AC motors is their constant high rotary speed (often measured in thousands of rounds per minute), which makes them useful in a wide range of applications like railway traction mechanisms, high-power engineering, commercial drill machines, industrial robots, and home appliances. The good performance characteristics of universal motors led to instant success in its widespread applications ranging from household appliances to sophisticated industrial equipment.

A universal motor is essentially the same as a direct current (DC) series motor. In a universal motor, wire is wrapped around the armature and the field in series, which means that the same current is applied to both sets of wire. As current passes through the field winding, a large magnetic field is generated. This field passes through the metal of the field winding, across an air gap between the field and the armature, through the armature winding, through the shaft of the armature, across another air gap, and back into metal of field winding, thus completing a magnetic circuit. When the magnetic field passes though the armature winding, the magnetic fields exerts a force in the current carrying wires. Because of the geometry of the windings, current on one side of the armature is always passing in the opposite direction to the current on the other side of the armature. Thus, the force exerted by the magnetic field on one side of the armature is opposite to the force exerted on the other side of the armature. Thereby a net torque is exerted on the armature, causing the armature to spin.

The performance of a universal motor is affected by several parameters such as radius of the motor (r_o), number of wire turns on each pole (N_s), thickness of the stator (t), length of air gap (l_{gap}), armature diameter (l_r), mean path length within the stator (l_c) etc. as shown in Fig. 7.2

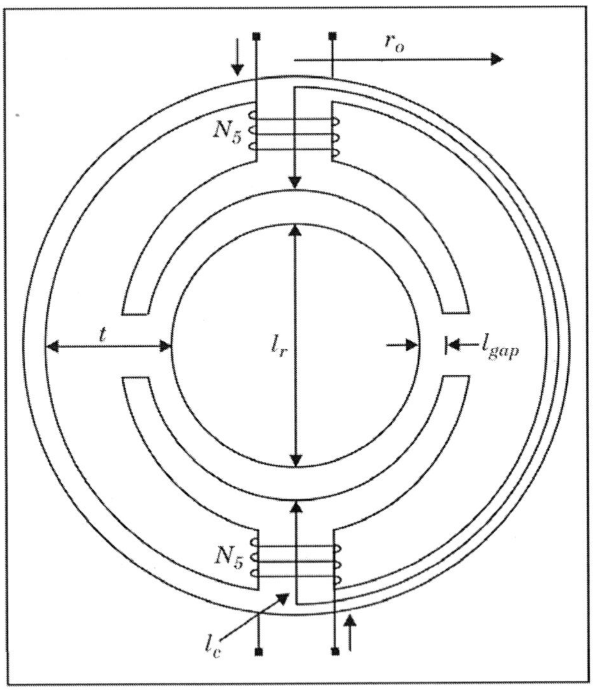

Fig. 7.2: Model geometry of a universal motor (Simpson *et al.*, 2001).

The optimization model for universal motor considered in the present work is based on the analysis given by Simpson *et al.* (2001). The eight decision variables considered for this model are, number of wire turns on the armature (N_c), number of wire turns on each pole of the field (N_s), cross sectional area of the wire on armature (A_{wa}), cross sectional area of the wire on the field (A_{wf}), radius of the motor (r_o), thickness of the stator (t_s), current drawn by the motor (I), and stack length (L). The two objectives considered in this work are discussed below.

Objectives

The first objective is to maximize the efficiency of the motor (h).The equation for efficiency can be computed from the equation of power. The basic equation for efficiency is given by:

$$\eta = \frac{P}{P_{in}} \tag{7.4}$$

$$\text{Where, } P = P_{in} - P_{out} \tag{7.5}$$

$$\text{Input power } P_{in} = V_t . I \tag{7.6}$$

V_t is the terminal voltage.

Power loss due to copper and brushes (P_{out}) is expressed as:

$$P_{out} = I^2 (R_a + R_s) + 2I \tag{7.7}$$

Where, R_a is the armature wire resistance and is given as

$$R_a = \frac{\rho l_{wa}}{A_{wa}} \times 10^6 \tag{7.8}$$

ρ is the resistivity of copper. l_{wa} is the armature wire length given by equation (7.9).

$$l_{wa} = (2L + 2l_r) N_c \tag{7.9}$$

$$l_r = 2(r_o - t_s - l_g) \tag{7.10}$$

Where, l_g is the length of air gap.

The stator wire resistance R_s is given as

$$R_s = \frac{\rho l_{wf}}{A_{wf}} \times 10^6 \tag{7.11}$$

Where, l_{wf} is the stator wire length given by equation (7.12)

$$l_{wf} = p_{st}[2L + 4(r_o - t_s)]N_s \tag{7.12}$$

P_{st} is the number of stator poles.

The second objective is to minimize the mass of the motor (M). To estimate the mass of motor, the motor is modeled as a solid cylinder and stator is modeled as a hollow steel cylinder. By including the mass of the winding on armature and the field, the equation for the mass (M) is given as:

$$M = M_s + M_a + M_w \tag{7.13}$$

Where, M_s is the mass of the stator given by equation (7.14)

$$M_s = \pi L(r_o^2 - (r_o - t_s)^2)\delta_s \tag{7.14}$$

δ_s is the density of steel.

M_a is the mass of the armature given by equation (7.15)

$$M_a = \pi L(r_o - t_s - l_g)^2 \delta_s \tag{7.15}$$

M_w is the mass of the winding given by equation (7.16)

$$M_w = (l_{wa}A_{wa} + l_{sw}A_{wf})\delta_c \times 10^{-6} \tag{7.16}$$

δ_c is the density of copper.

Constraints

The objectives given by Eq. (7.4) and Eq. (7.13) are subjected to following constraints:

1. *Magnetizing intensity*

The constraint on magnetizing intensity (H) ensures that magnetic flux within each motor does not exceed the physical flux carrying capacity of steel. This constraint is specified by Eq. (7.17).

$$H \leq 5000\,\text{A turns/m} \tag{7.17}$$

H is given by Eq. (7.18).

$$H = \frac{N_c I}{(l_c + l_r + 2l_g)} \tag{7.18}$$

The mean path length within stator (l_c) and the diameter of the armature (l_r) are obtained by equations (7.19) and (7.20) respectively.

$$l_c = \frac{\pi(2r_o + t_s)}{2} \tag{7.19}$$

$$l_r = 2(r_o - t_s - l_g) \tag{7.20}$$

2. *Feasible geometry*

The constraint on feasible geometry ensures that the stator thickness does not exceed radius of stator since the thickness is measured from the outside of the motor inward. This constraint is given as

$$t_s < r_o \tag{7.21}$$

Specific customer attributes:

The motor should be designed to satisfy following specific customer attributes.

1. *Torque (T):*

$$T = K\phi I \tag{7.22}$$

Where, K = motor constant $= \dfrac{N_c}{\pi}$

ϕ is magnetic flux given by equation (7.23).

$$\phi = \frac{\mathfrak{I}}{\mathfrak{R}} \tag{7.23}$$

$\Im$ is the magneto magnetic force expressed as:

$$\Im = N_s . I \tag{7.24}$$

$\Re$ is the total reluctance given by equation (7.25).

$$\Re = \Re_s + \Re_a + 2\Re_g \tag{7.25}$$

$$\text{Stator reluctance } \Re_s = \frac{l_c}{2\mu_{steel}\mu_0 A_s} \tag{7.26}$$

$$\text{Armature reluctance } \Re_a = \frac{l_r}{\mu_{steel}\mu_0 A_a} \tag{7.27}$$

$$\text{Reluctance of air gap } \Re_g = \frac{l_c}{\mu_0 A_g} \tag{7.28}$$

Cross sectional area of stator (A_s) is given as:

$$A_s = t_s . L \tag{7.29}$$

Cross sectional area of armature (A_a) is given as:

$$A_a = l_r . L \tag{7.30}$$

Cross sectional area of air gap (A_g) is given as:

$$A_g = l_g . L \tag{7.31}$$

If $H \leq 220$

$$\mu_{steel} = -0.2279\, H^2 + 52.411\, H + 3115.8 \tag{7.32}$$

If $220 < H \leq 1000$

$$\mu_{steel} = 11633.5 - 1486.33 \ln (H) \tag{7.33}$$

If $H > 1000$

$$\mu_{steel} = 1000 \tag{7.34}$$

2. *Power (P)*

To summarize, the optimization model thus consist of two objectives: (i) To maximize the efficiency and (ii) To minimize the mass of the motor as given by equations (7.4) and (7.13) respectively. The constraints on magnetizing intensity and the feasible geometry are given by equations (7.17) and (7.21) respectively. Moreover, the specific requirements of the customer for torque of 0.05 Nm and power of 300 W must also be satisfied. The parameters bounds considered as:

$100 \le N_c \le 1500$ turns; $1 \le N_s \le 500$ turns; $0.01 \le A_{wa} \le 1$ mm^2; $0.01 \le A_{wf} \le 1$ mm^2, $0.01 \le r_o \le 0.1$ m; $0.0005 \le t_s \le 0.01$m; $0.1 \le I \le 6$ A; $0.01 \le L \le 0.2$ m

Values of constants used in the present example are length of the air gap (L_g) =7×10^{-4}m, Terminal voltage (V_T) = 115 V, Resistivity of copper (ρ) = 1.69×10^{-8}Wm, Permeability of free space (μ_0) = 4p$\times10^{-7}$H/m, Number of stator poles = 2.

The above multi-objective optimization problem is now considered to demonstrate the working of SA algorithm. The normalized combined objective function (*Z*) is formulated considering different weightages to all objectives and is given by the following Eq. (7.33).

$$\text{Minimize } Z = -\frac{w_1 \times \eta}{\eta_{max}} + \frac{w_2 \times M}{M_{min}} - P_i \times CV_i \tag{7.33}$$

Where, η_{max} = maximum value of efficiency obtained when single objective optimization problem considering only minimization of efficiency as an objective, was solved for the given constraints = 0.938.

M_{min} = minimum value of mass of motor when single objective optimization problem considering only minimization of mass as an objective was solved for the given constraints = 0.141.

P_i = Penalty for i^{th} constraint and CV_i = Amount of constraint violation for i^{th} constraint w_1 and w_2 are the weights assigned to the objective functions η and M respectively. The values of weightages can be calculated by using the analytic hierarchy process (AHP). However, in the present example, equal weights are considered for both the objectives *i.e.,* w_1 = 0.50, w_2 = 0.50.

Step 1: ***Set initial conditions***

The following parameters of optimization are selected for SA algorithm.

- Initial temperature = 200
- Decrement factor = 0.01
- Termination temperature = 10

Initial Solution: N_c= 129.5, N_s = 167.9, A_{wa} = 0.344, A_{wf} = 0.747, r_o = 0.061, t_s = 0.03, I = 1.324, L = 0.134 with Z = 586.85

Step 2: ***Generate new solution***

Considering five random numbers (N=5) and the standard deviations for the variables as σ_{NC} = 233.33, σ_{NS} = 83.166, σ_{Awa} = 0.165, σ_{awf} = 0.655, σ_{ro} = 0.015, σ_{ts} = 0.00158, σ_{I} = 0.833, σ_{L} = 0.0316, using Eq. (7.2), the new solution generated in the vicinity of initial solution is: N_c= 171.13, N_s= 262.55, A_{wa}= 0.370, A_{wf}= 0.920, r_o= 0.056, t_s= 0.002, I = 1.576, L = 0.171 with Z' = 347.47.

Step 3: ***Apply rules of acceptance of new solution***

Since new solution is feasible and superior to initial solution, is accepted by rule 1 in Eq. (7.3). Thus the move is successful and the temperature is reduced to 198°C.

Step 4: ***Check whether termination criteria is met***

Since the termination criteria are not met, the steps 2 to 4 are repeated until the temperature goes below T_e.

Results obtained for next five iterations are shown in Table 7.1.

It is to be noted that solutions obtained in iterations 2, 3 and 5 are superior to their previous values, hence selected with 100% probability and the temperature will be reduced. However, the solutions 4 and 6 are inferior to their respective previous solutions and hence are accepted with probabilities 90% and 54% according to the rule 2 of Eq. (7.3). Out of which the solution 4 gets accepted and hence the temperature is reduced in iteration 4, however, solution 6 gets rejected and hence the temperature will remains same and new solution will be generated in the vicinity of solution 5 with new set of random numbers. The iterations are continued till the termination temperature is reached.

Table 7.1: Results obtained by running SA algorithm.

It. no	*Variables*								
	N_c	N_s	A_{wa}	A_{wf}	r_o	t_s	I	L	Z
2	171.134	262.551	0.370	0.920	0.056	0.002	1.576	0.171	347.41
3	212.729	357.169	0.396	1.000	0.052	0.001	1.828	0.200	331.72
4	237.546	319.587	0.317	0.826	0.044	0.001	1.677	0.213	352.22
5	158.681	348.372	0.119	0.869	0.026	0.002	2.223	0.179	278.54
6	205.520	294.782	0.101	0.724	0.014	0.003	1.841	0.157	397.14

Table 7.1: Continued...

It. no.	Z	ΔObj	*Pr*	*RN*	*Remark*	T
2	347.41	–239.3	---	---	Accepted	198.00
3	331.72	–15.72	---	---	Accepted	196.02
4	352.22	20.46	0.90	0.83	Accepted	194.05
5	278.54	–73.68	---	---	Accepted	192.11
6	397.14	118.6	0.54	0.68	Rejected	192.11

The results of optimization for improving the performance of universal motor in terms of maximizing efficiency and minimizing weight are:

$N_c = 100.5$, $N_s = 500$, $A_{wa} = 0.123$, $A_{wf} = 0.121$, $r_o = 0.0173$, $t_s = 0.00645$, $I = 4.321$, $L = 0.01$ with $T = 0.05$ N-m, $P = 307.3$ W, $h = 61.8\%$, $M = 0.145$ Kg, and $Z = 0.1872$.

7.4. Modelling and Optimization of Process Parameters of Injection Molding Process

Plastics are now replacing metals in many applications as they offer wide variety of properties particularly fitted for modern developments. Among various plastic molding processes, injection molding is one of the most widely accepted and important methods applied for forming plastic products in industry due to short product cycles, high quality part surfaces, good mechanical properties, low cost, and light weight. Injection molding can process not only thermoplastics but thermosetting plastics also. The injection molding process involves the injection of polymer melt into a mold where the melt cools and solidifies to form a plastic product. It is generally a three phase process comprising filling, packing, and cooling phases. Injection molding machine users always want to achieve higher productivity with a desired accuracy and quality of plastic product. Performance of the injection molding process, however, is affected by many factors such as melt temperature, mold temperature, injection pressure, injection velocity, injection time, packing pressure, packing time, cooling temperature, cooling time etc. and a single parameter change will influence the process in a complex way.

Cycle time, percent mold filling, and mold temperature are considered to be important performance measures for plastic injection molding process, hence the same are considered in this case study. A mathematical model relating these performance measures with four important process parameters namely, Injection speed setting (V_i), Melt temperature (T_m), Injection time (T_i), and Injection pressure (P_i) is developed using response surface methodology and the optimum parameter setting is achieved through simulated annealing algorithm.

The coded value corresponding to actual value for each process variable is derived using following formula:

$$\text{Coded test condition} = \frac{\text{Actual test condition - mean test condition}}{\text{Range of test condition/2}}$$

The coded numbers are thus obtained from following transformation equation:

$$x_1 = \frac{V_i - V_{i_0}}{\Delta V_i} \tag{7.34}$$

$$x_2 = \frac{T_m - T_{m_0}}{\Delta T_m} \tag{7.35}$$

$$x_3 = \frac{T_i - T_{i_0}}{\Delta T_i} \tag{7.36}$$

$$x_4 = \frac{P_i - P_{i_0}}{\Delta P_i} \tag{7.37}$$

Where, x_1, x_2, x_3 and x_4 are the coded values of the variables V_i, T_m, T_i, and P_i respectively. V_{i0}, T_{m0}, T_{i0}, and P_{i0} are the values of injection speed setting, melt temperature, injection time, and injection pressure at zero level. ΔV_i, ΔT_m, ΔT_i, and ΔP_i are the intervals of variation in V_i, T_m, T_i, and P_i respectively. Table 7.2 shows coded values of process variables.

Table 7.2: Coded values of process variables for case study 7.2.

Factors	***Coded levels***				
	–2	***–1***	***0***	***+1***	***+2***
Injection speed setting (x_1)	50	60	70	80	90
Melt temperature (x_2)	200	210	220	230	240
Injection time (x_3)	0.75	1	1.25	1.5	1.75
Injection pressure (x_4)	40	70	100	130	160

The details of experimental set up used for data collection are given below:

- Work material: Poly propylene
- Machine type/make: Optima series, model (075#10) S.R. No. 483.
- Work-piece specification: L-shape spacer of size 30×30×20 mm and thickness 5 mm.
- Number of impressions: 8

The experimental matrix adopted in the present study in the coded form is shown in Table 7.3.

To study the effect of process parameters *i.e.*, V_i, T_m, T_i, and P_i, on performance measures *i.e.*, cycle time (T_c) and mold temperature (T_{mold}) and percent filling (F_p), a second-order polynomial response is fitted into equation (3.29). Equations (7.38), (7.39) and (7.40) are then derived by determining the values of the coefficients using the least square technique for data collected as shown in Table 7.3, cycle time (T_c), percent filling (F_p) and mold temperature (T_{mold}) respectively.

$$T_c = 19.43 + 0.1458x_1 - 0.1708x_2 + 0.0391x_3 - 0.1275x_4 - 0.1987x_1x_2 + 0.05625x_1x_3 - 0.07625x_1x_4 - 0.01375x_2x_3 + 0.03875x_2x_4 + 0.2137x_3x_4 + 0.3723x_1^2 + 0.0873x_2^2 - \quad (7.38)$$

$$0.0377x_3^2 - 0.06521x_4^2$$

$$F_p = 96.25 + 3.333x_1 + 1.6667x_2 + 15x_3 + 3.75x_1x_2 - 5x_1x_3 + 3.75x_1x_4 - 2.5x_2x_3 - 1.25x_2x_4 - 0.729x_1^2 - 0.729x_2^2 - 6.979x_3^2 - 0.729x_4^2 \quad (7.39)$$

$$T_{mold} = 30.25 + 0.04166x_1 - 0.375x_2 + 0.2083x_3 + 0.625x_4 - 0.4375x_1x_2 - 0.1875x_1x_3 - 0.6875x_1x_4 + 0.6875x_2x_3 - 0.5625x_2x_4 + 0.4375x_3x_4 - 0.01042x_1^2 + 0.4895x_2^2 + 0.8645x_3^2 + 0.9895x_4^2 \quad (7.40)$$

Table 7.3: Experimental data collection (Pawar, 2010).

x_1	x_2	x_3	x_4	T_{mold} (°C)	T_c (Sec)	F_p (%)
–1	–1	–1	–1	31.33	19.92	65.83
1	–1	–1	–1	34.04	20.65	67.50
–1	1	–1	–1	31.21	19.93	69.17
1	1	–1	–1	32.17	19.86	85.83
–1	–1	1	–1	29.87	19.49	110.83
1	–1	1	–1	31.83	20.44	92.50
–1	1	1	–1	32.50	19.44	104.17
1	1	1	–1	32.71	19.60	100.83
–1	–1	–1	1	34.21	19.31	60.83
1	–1	–1	1	34.17	19.74	77.50
–1	1	–1	1	31.83	19.47	59.17
1	1	–1	1	30.04	19.10	90.83
–1	–1	1	1	34.50	19.73	105.83
1	–1	1	1	33.71	20.38	102.50
–1	1	1	1	34.87	19.84	94.17
1	1	1	1	32.33	19.69	105.83
0	0	0	0	30.25	19.43	96.25
0	0	0	0	30.25	19.43	96.25
0	0	0	0	30.25	19.43	96.25
0	0	0	0	30.25	19.43	96.25
2	0	0	0	30.29	21.21	100.00
–2	0	0	0	30.13	20.63	86.67
0	2	0	0	31.46	19.44	96.67
0	–2	0	0	32.96	20.12	90.00
0	0	2	0	34.12	19.36	98.33
0	0	–2	0	33.29	19.20	38.33
0	0	0	2	35.46	18.91	93.33
0	0	0	–2	32.96	19.42	93.33

To test whether the data are well fitted in model or not, the values of standard error of estimates (S) of the regression analysis for cycle time, filling percent, and mold temperature are obtained as 0.55, 10.80 and 2 respectively, whereas the values of standard deviation (S_y) for cycle time, filling percent, and mold temperature are obtained as 0.93, 26.96, and 3.17 respectively. As $S < S_y$ for regression analysis of all the three responses, indicates that regression models obtained for cycle time, filling percent, and mold temperature have merit. The actual extent of improvement, by using regression analysis rather than describing data as an average value, is quantified by coefficient of determination (R^2), which varies from 0-1. Value of $R^2 = 1$ indicates perfect fit, whereas $R^2 = 0$ indicates no improvement. For cycle time, filling percent, and mold temperature, R^2 values are calculated as 0.67, 0.84 and 0.60 respectively. The R^2 value is moderately high for filling percent and is moderate for cycle time and mold temperature. Hence, the models developed for cycle time, filling percent, and mold temperature fits the data well.

Now to demonstrate and validate the SA algorithm, an example is considered for the optimization of injection molding process parameters for an optimization model discussed below:

Objective function

Minimize T_c (given by Eq. 7.38)

Constraints

1. *Filling constraint*

The parameter selected should be such that all impressions are filled completely. This constraint is as given below.

$$F_P - 100 \geq 0 \tag{7.41}$$

2. *Mold temperature constraint*

The ability to control the mold temperature is an important factor in plastic production. It is an effective tool for maintaining minimum cycles and controlling dimensions and mechanical properties. For most materials the uniformity in appearance is largely dependent on the consistency of mold surface temperature. For the component considered in this case study, the mold temperature should not fall below 34°C. This constraint is specified as:

$$T_{mold} - 34 \geq 0 \tag{7.42}$$

Considering the objective function and constraints together, the combined objective function (z) is formulated as:

$$z = T_c - CV_1 \times P_1 - CV_2 \times P_2 \tag{7.43}$$

Where, CV_1 is amount of constraint violation for constraint given by Eq. (7.41) and CV_2 is amount of constraint violation for constraint given by Eq. (7.42). P_1 and P_2 are the static penalties assigned to constraints given by Eq. (7.41) and Eq. (7.42) respectively.

Parameters and parameter bounds

The four process parameters considered in the present work are, Injection speed setting (V_i), Melt temperature (T_m), Injection time (T_i), and Injection pressure (P_i) .The upper and lower bound values for these parameters are as given below.

$50 \leq V_i \leq 90\%$; $0 \leq T_m \leq 240°C$; $0.75 \leq T_i \leq 1.75$ sec; $40 \leq P_i \leq 160$ bar

These parameters when expressed in their coded form, the bounds are $-2 \leq x_i \leq 2$, for $i = 1$ to 4.

Now, the steps of SA algorithm are applied as discussed below:

Step 1: ***Set initial conditions***

The following parameters of optimization are selected for SA algorithm.

- Initial temperature = 500
- Decrement factor = 0.01
- Termination temperature = 20

Initial Solution (in coded values): $x_1 = -1.615$, $x_2 = 0.898$, $x_3 = 0.484$, $x_4 = -0.708$ with z = 361.02.

Step 2: ***Generate new solution***

Considering five random numbers (N=5) and the standard deviations for all variables as 0.667, using Eq. (7.2), the new solution generated in the vicinity of initial solution is:

x_1= -1.825, x_2= 0.903, x_3 = 0.315, x_4 = -0.557 with z = 478.76.

Step 3: ***Apply rules of acceptance of new solution***

It is thus observed that the new solution is feasible but inferior to initial solution. Hence its probability of acceptance is evaluated by rule 2 of Eq. (7.4) and is 0.704. Let the random number generated is 0.84, so the new solution is rejected and the temperature remains unchanged. Another solution is now generated with some other set of random numbers and is:

x_1= –1.572, x_2= 0.798, x_3 = 0.922, x_4 = –1.174 with z = 235.92

The new solution is now superior to initial solution and is feasible also. Hence it is accepted by rule 1 in Eq. (7.3). Thus the move is successful and the temperature is reduced to 495°C. It is to be noted that combined objective function of this updated solution is superior to initial solution and it also satisfies the constraint of % filling as against the initial solution for which both constraints are violated.

Step 4: ***Check whether termination criteria is met***

Since the termination criteria are not met, the steps 2 to 4 are repeated until the temperature goes below T_e.

Table 7.4 shows the results obtained for next five generations.

Table 7.4: Results of parametric optimization of injection molding process.

It.	*Variables*				*Z*	*ΔObj*	P_r	*RN*	*Remark*	*T*
no.	x_1	x_2	x_3	x_4						
2	–1.981	0.867	1.353	–1.069	121.28	–114.64	–	–	Accepted	490.05
3	–1.873	0.920	1.043	–1.561	153.68	32.40	0.93	0.56	Accepted	485.14
4	–1.256	0.295	0.530	–0.845	394.37	240.63	0.60	0.94	Rejected	485.14
5	–1.909	0.561	1.473	–1.310	164.59	10.91	0.97	0.12	Accepted	480.28
6	–2.000	0.193	1.907	–1.379	132.47	–32.23	–	–	Accepted	475.47

It is to be noted that solution obtained in iteration 2 is superior to its previous value, hence accepted with 100% probability. As the move is successful, the temperature will be reduced. However, the solution obtained in iteration 3 is inferior to its previous solutions and hence is accepted with probability of 89% according to the rule 2 of Eq. (7.3).The iterations are continued till the termination temperature is reached. Solution obtained in iteration 4 is also inferior to solution 3 and has probability of acceptance 64%. However, since this probability is less than random number generated, this solution is rejected and temperature remains the same. The solution 5 is then generated from solution 3 (and not from solution 4). Solution in generation 6 is superior to solution of previous iterations is accepted. The results of optimization to minimize the cycle time of injection molding process are:

$x_1 = -0.531$, $x_2 = 0.064$, $x_3 = 2.000$, $x_4 = -2.000$ with z = 18.37.

The corresponding decoded values are: $V_i = 64.69\%$, $T_m = 220.64°C$, $T_i = 1.75$ Sec, $P_i = 40$ bar. The cycle time (T_c) is 18.37 sec. and both constraints are well satisfied.

7.5. Variants of Simulated Annealing Algorithm

Several variants of simulated annealing algorithm are proposed by the researchers to improve its performance as discussed below:

7.5.1. *Simulated Annealing with non-monotonic reheating*

In simulated annealing algorithm, at early stages of evolution, the probability of acceptance of an inferior solution is quite large. However, as the temperature reduces with eventual successful moves, the solution with same amount of inferiority will not be accepted. However, even though the solution is inferior, it may have a potential to search global optimum solution. However, it gets rejected due to small temperature and no matter how slow the cooling rate, from that point on, the algorithm will always deliver a local minimum. The only solution to this problem is to detect that a local minimum has been reached and to reheat the system, thus allowing it to escape from local minimum. There are various reheating schemes that have been proposed. Nguyen and Duong (2009) used the reheating scheme called non-monotonic reheating. The basic idea of non-monotonic reheating is that whenever the occurrence of a local minima is detected, this scheme attempts to escape from it by resetting the temperature to twice of the temperature at which the best solution was attained. This scheme starts from an initial temperature, which is decremented after each Markov chain according to a standard temperature decrement rule. When, at an iteration k, a local minimum is detected (a cycle of search is made in which no transition is accepted), the temperature at the iteration $k+1$, T_{k+1}, is updated according to the following rule: $T_{k+1} = 2T_b$. Here, T_b is the temperature corresponding to the best solution obtained. The algorithm stops when a certain number of reheats have been performed.

7.5.2. *Very fast simulated re-annealing*

The standard simulated annealing needs a lot of time to converge to the optimal solution. It is well-known that the standard SA requires a cooling schedule in which the temperature T must satisfy Eq. (7.44).

$$T_k = T_0 / \ln (1+k) \tag{7.44}$$

Where T_k is the temperature at the k-th iteration and T_0 is the initial temperature.

Equation (7.44) characterizes the simplest cooling scheme for SA which is also called Boltzmann annealing. Since k must become very large for T_k approaching to zero, the equation can explain why this cooling scheme is always slow. To overcome this issue, Szu and Hartley (1987) proposed the use of Cauchy distribution to find global optimum. With this scheme the Eq. (7.44) is replaced by Eq. (7.45).

$$T_k = T_0 / k \tag{7.45}$$

This method is called as fast annealing (FA) and is statistically seen to have an annealing schedule exponentially faster than the method of basic SA algorithm.

In a D – dimensional parameter space, every parameter has its own range fixed by physical consideration. For different ranges there might be different annealing-time-dependent sensitivities, measured by the curvature of the cost-functional local minima. However, both standard SA and fast SA algorithms assumes infinite ranges for all parameters and there is no provision for considering differences in each parameter-dimension, *e.g.,*, different sensitivities might require different annealing schedules. Also, there is no quick algorithm for calculating a D-dimensional Cauchy random generator. To overcome this issue, Ingber and Rosen (1992) proposed a new formula to calculate the annealing schedule considering differences in each parameter dimension as:

$$T_i(k) = T_{oi} \times e^{(-c_i.k^{1/D})} \tag{7.46}$$

Where, $T_i(k)$ is temperature of ith parameter at k^{th} iteration, a T_{oi} is the initial temperature of ith parameter, c_i is the constant, and D is number of parameters.

The very fast simulated re-annealing algorithm was later renamed as adaptive simulated annealing algorithm.

7.5.3. *Hybrid PSO-SA*

The Particle Swarm-Based-Simulated Annealing (PSO-B-SA) proposed by Sadati *et al.* (2009) is an optimization algorithm which combines the PSO with the SA. In fact by combining PSO with SA, the strong points of SA can be used in PSO. This is the basic idea of the PSO-B-SA. The PSO-B-SA algorithm's searching process is started from initializing a group of random particles. In this process, either only *gBest* which is the leader of the swarm is based on SA, independently from other particles, or each particle is based on SA. In the process of simulated annealing, the new individuals x(t+1) can be generated randomly around the original ones x(t) as given by Eq. (7.2).

Now, to find the global minimum of an optimizing problem, the steps taken in the Particle Swarm-Based-Simulated Annealing Optimization is as follows:

(1) Initialize a group of particles (the number of particles is m), including random position and velocity.
(2) Evaluate each particle's fitness.
(3) *gBest* is based on SA independently, and a new global best position (*gBest*) is obtained.
(4) For each particle, compare its fitness and its personal best position (*pBest*). If its fitness is better, replace *pBest* with its fitness.
(5) For each particle, compare its fitness and the global best position (*gBest*). If its fitness is better, then replace *gBest* with its fitness.

(6) Transform each particle's velocity and its position according to the expressions given by equations (3.1) and (3.2) respectively.

(7) This process evolves through time until the terminating condition is satisfied.

It is reported that the hybrid Particle Swarm-Based-Simulated Annealing Optimization algorithm can identify a global optimum solution in minimum runs, in compare to other optimization techniques.

8

Teaching Learning Based Optimization

8.1. Introduction

Teaching-learning-based optimization (TLBO) algorithm is a teaching-learning process inspired algorithm proposed by Rao *et al.* (2011); Rao *et al.* (2012) and Rao and Patel (2012), which is based on the effect of influence of a teacher on the output of learners in a class. The algorithm mimics the teaching-learning ability of teacher and learners in a class room. Teacher and learners are the two vital components of the algorithm and describes two basic modes of the learning, through teacher (known as teacher phase) and interacting with the other learners (known as learner phase). The output in TLBO algorithm is considered in terms of results or grades of the learners which depend on the quality of teacher. So, teacher is usually considered as a highly learned person who trains learners so that they can have better results in terms of their marks or grades. Moreover, learners also learn from the interaction among themselves which also helps in improving their results. TLBO is population based method. In this optimization algorithm a group of learners is considered as population and different design variables are considered as different subjects offered to the learners and learners' result is analogous to the 'fitness' value of the optimization problem. In the entire population the best solution is considered as the teacher.

The working of TLBO is divided into two parts, 'Teacher phase' and 'Learner phase'. Working of both the phases is explained below.

8.1.1. *Teacher phase*

During this phase a teacher tries to increase the mean result of the class in the subject taught by him or her depending on his or her capability. At any iteration *i*, assume that there are '*m*' number of subjects (*i.e.,* design variables), 'n' number of learners (*i.e.,* population size, k=1,2,...,n) and $M_{j,i}$ be the mean result of the learners in a particular subject '*j*' (j=1,2,...,m) The best overall result $X_{total\text{-}kbest,i}$ considering all the subjects together obtained in the entire population of learners can be considered as the result of best learner *kbest*. However, as the teacher is usually considered as a highly learned person who trains learners so that they can have better results, the best learner identified is considered by the algorithm as the teacher. The difference between the existing mean result of each subject and the corresponding result of the teacher for each subject is given by,

$$Difference\_Mean_{j,k,i} = r_i \times (X_{j,kbest,i} - T_F \times M_{j,i}) \quad (8.1)$$

Where, $X_{j,kbest,I}$ is the result of the best learner (*i.e.,* teacher) in subject *j*. T_F is the teaching factor which decides the value of mean to be changed, and r_i is the random number in the range [0, 1]. Value of T_F can be either 1 or 2. The value of T_F is decided randomly with equal probability as,

$$T_F = round\ [1+rand\ (0,1)\{2-1\}] \quad (8.2)$$

T_F is not a parameter of the TLBO algorithm. The value of T_F is not given as an input to the algorithm and its value is randomly decided by the algorithm using equation (8.2). After conducting a number of experiments on many benchmark functions it is concluded that the algorithm performs better if the value of T_F is between 1 and 2. However, the algorithm is found to perform much better if the value of T_F is either 1 or 2 and hence to simplify the algorithm, the teaching factor is suggested to take either 1 or 2 depending on the rounding up criteria given by equation (8.2).

Based on the *Difference_Mean*$_{j,k,i}$, the existing solution is updated in the teacher phase according to the following expression.

$$X'_{j,k,i} = X_{j,k,i} + Difference\_Mean_{j,k,i} \quad (8.3)$$

Where $X'_{j,k,i}$ is the updated value of $X_{j,k,i}$. Accept $X'_{j,k,I}$ if it gives better function value. All the accepted function values at the end of the teacher phase are maintained and these values become the input to the learner phase. The learner phase depends upon the teacher phase.

8.1.2. *Learner phase*

Learners increase their knowledge by interaction among themselves. A learner interacts randomly with other learners for enhancing his or her knowledge. A learner learns new things if the other learner has more knowledge than him or her. Considering a population size of '*n*', the learning phenomenon of this phase is expressed below.

Randomly select two learners P and Q such that $X'_{total-P,i} \neq X'_{total-Q,i}$ (where, $X'_{total-P,i}$ and $X'_{total-Q,I}$ are the updated values of $X_{total-P,i}$ and $X_{total-Q,i}$ respectively at the end of teacher phase)

$$X''_{j,P,i} = X'_{j,P,i} + r_i\,(X'_{j,P,i} - X'_{j,Q,i}),\ \text{If } X'_{total-P,i} < X'_{total-Q,i} \tag{8.4}$$

$$X''_{j,P,i} = X'_{j,P,i} + r_i\,(X'_{j,Q,i} - X'_{j,P,i}),\ \text{If } X'_{total-Q,I} < X'_{total-P,i} \tag{8.5}$$

Accept $X''_{j,P,i}$ if it gives a better function value.

The flow chart of TLBO algorithm is shown in Fig. 8.1.

8.2. Mechanism of Working of Teaching Learning Based Optimization Algorithm

The mechanism of Teaching Learning Based Optimization algorithm is explained through example: $Min. z = x_1^2 - 3x_2 + 10$ such that $0 \leq x_{1,}x_2 \leq 3$.

Step 1: ***Set initial conditions***

Since TLBO algorithm is claimed to have no 'algorithm specific parameters', the only parameter one needs to define is the population size *i.e.,* number of students (*N*) in class.

Let $N = 10$.

Step 2: ***Generate initial population***

In TLBO, the number of subjects represents the number of parameters and parameter value is represented by marks in that subject. The objective function value is represented by the score of the student. For

10 students the marks in each subject and scores are as shown in Table 8.1.

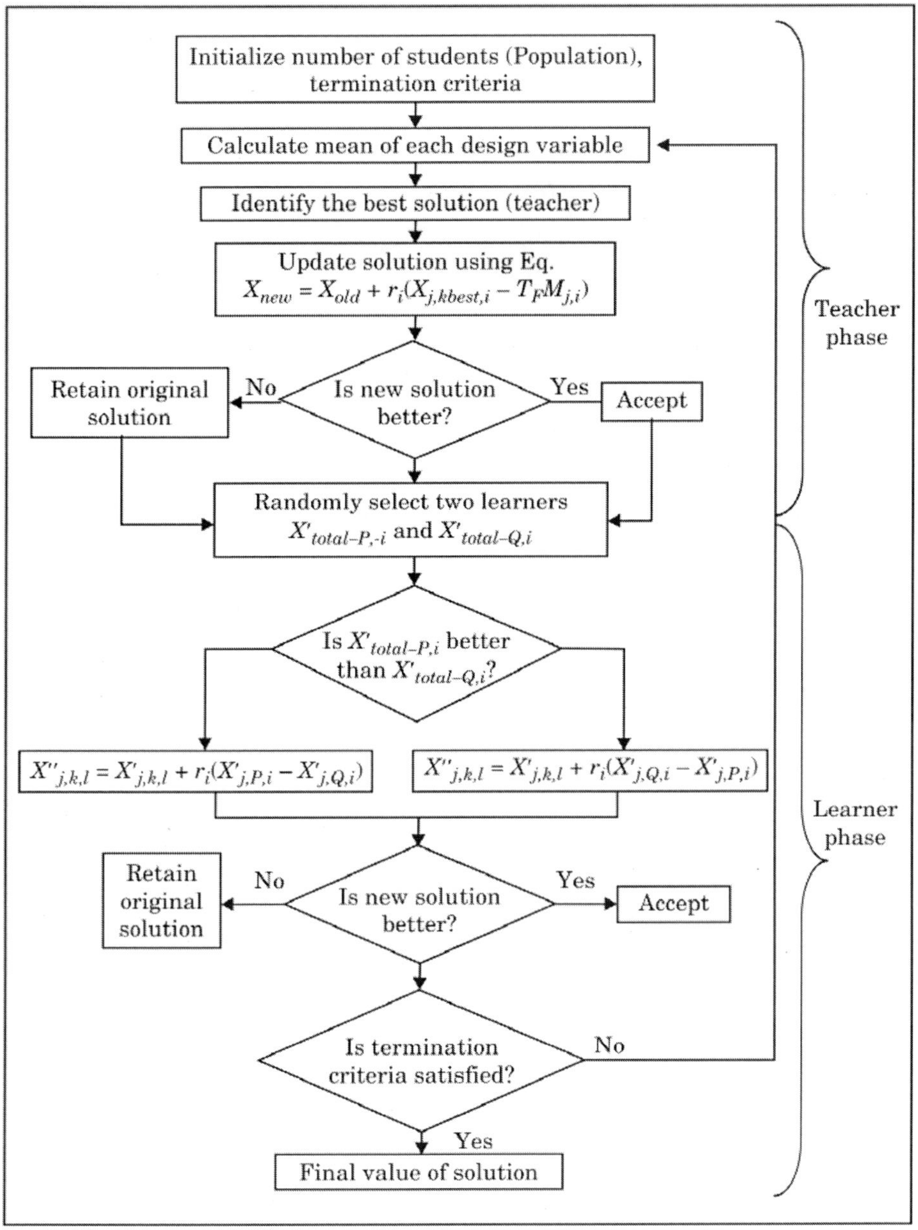

Fig. 8.1: Flowchart of TLBO algorithm (Rao *et al.*, 2011).

Table 8.1: Initial population of solution.

Student roll number (solution number)	*1*	*2*	*3*	*4*	*5*	*6*	*7*	*8*	*9*	*10*
Marks in subject 1 (x_1)	1.43	1.47	1.51	1.79	2.60	2.35	1.11	0.18	1.38	0.65
Marks in subject 2 (x_2)	2.72	2.17	0.23	2.02	0.03	1.57	0.81	0.94	0.04	2.64
Score (z)	3.87	5.64	11.62	7.15	16.67	10.79	8.80	7.21	11.78	2.51

Step 3: *Improve score in teacher phase*

In teacher phase students learn from teacher. A teacher is considered to be the most knowledgeable and hence represented by the global best solution in population. Since the problem is of minimization the best solution in Table 8.1 is solution number 10 with score 2.51. Each student will learn from the teacher to improve score as given by Eq. (8.1) to Eq. (8.3). The updated marks and new scores of the students are shown in Table 8.2.

Table 8.2: Scores of students after implementing teacher phase.

Student roll number (solution number)	*1*	*2*	*3*	*4*	*5*	*6*	*7*	*8*	*9*	*10*
Updated marks in subject 1 (x_1')	0.7	0.9	1.15	1.1	2.54	2.1	0.8	0	0.96	0.3
Updated marks in subject 2 (x_2')	2.8	2.7	1.06	3	1.19	1.76	1.1	1.7	1.19	3
Score (z')	2.1	2.7	8.14	2.2	12.9	9.12	7.3	5	7.37	1.1

Step 4: *Improve scores in Learner phase*

In learner phase, the students learn by sharing their knowledge with each other and improve their scores. The student having higher score will share his knowledge with student having less score and the student under consideration will receive knowledge partly from this information exchange as given by Eq. (8.4) and Eq. (8.5). Updated marks and new scores of the students after implementing Lerner phase are shown in Table 8.3. It may be noted that solutions 4, 8, and 10 obtained in teacher phase are retained as their updated values in the learner phase are inferior.

Table 8.3: Scores of students after implementing learner phase.

Student roll number (solution number)	***1***	***2***	***3***	***4***	***5***	***6***	***7***	***8***	***9***	***10***
Updated marks in subject 1 (x_1")	0.4	0.1	1.17	1.1	1.22	2.03	0.5	0	0.74	0.3
Updated marks in subject 2 (x_2")	3	2.7	1.17	3	2.29	1.95	1.4	1.7	1.2	3
Score (z")	1.3	1.8	7.84	2.2	4.61	8.28	6.1	5	6.93	1.1

At the end of first generation, the teacher (*gBest*) performance is improved from z = 2.51 to z"= 1.1. Average fitness value is also improved from 8.60 in initial population to 4.51 after first iteration.

Step 5: *Repeat steps 2 to 4 till termination criteria is met*

The steps 2 to 4 are repeated till termination criteria is met.

8.3. Optimization of Cold Backward Extrusion Process

Extrusion is a bulk forming process in which the work metal is forced or compressed to flow through a die hole to produce a desired cross-sectional shape. This process offers several advantages including production of variety of shapes (usually by hot extrusion), achieving close tolerances and enhancement of grain structure and strength (usually by cold extrusion). This process can be classified as direct extrusion and indirect extrusion.

***Direct extrusion*:** A metal billet is first loaded into a container having die holes. A ram compresses the material, forcing it to flow through the die holes.

***Indirect (backward) extrusion*:** In this type, the die is mounted to the ram and not on the container. As the ram compresses the metal, it flows through the die hole on the ram side which is in opposite direction to the movement of ram. Since there is no relative motion between the billet and the container, there is no friction at the interface, and hence the ram force is lower than in direct extrusion.

The product considered in this case study is a collapsible tube of aluminum material used to store the medicinal cream. The details of this tube are shown in Fig. 8.2.

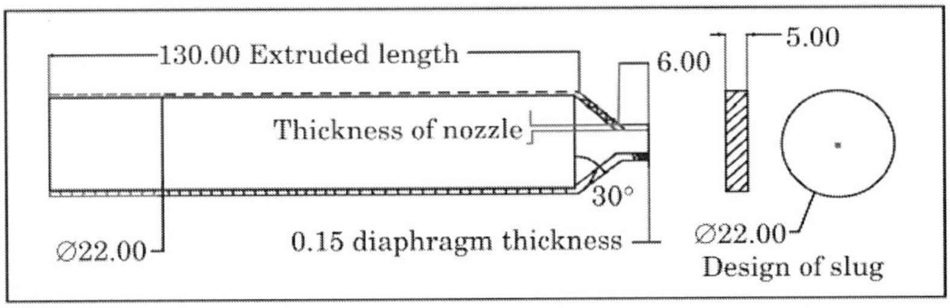

Fig. 8.2: Specification of collapsible tube to store the medicinal cream.

The tube is to be extruded from a slug having diameter 22 mm. The critical aspects of this tube manufacturing using cold backward extrusion process are to maintain the thickness of the diaphragm within the range of 0.10 to 0.15 mm, to maintain the thickness of the nozzle within the range of 1.15 mm to 1.525 mm, to ensure that the deviation of extruded length from its specified value (130 mm) should be minimum and the required extrusion power should be minimum. A horizontal type cold backward extrusion press of 100 ton is utilized to manufacture this tube. Keeping in view the above mentioned aspects of product manufacturing, the two objectives considered are (i) minimization of the deviation of tube length from its specified value of 130 mm. $L_d = |130 - L|$ and (ii) Minimization of the extrusion power required (P_w) subjected to the constraints of (i) thickness of diaphragm (t_d) to be maintained within 0.10 mm- 0.15 mm and (ii) thickness of nozzle (t_n) to be maintained within 1.15 mm- 1.525 mm. The important process variables affecting the above mentioned responses are extrusion force (F_e), stroke length (S_l), slug diameter (d_s), and back pressure (P_b). Operating ranges of these process variables and their coded levels are presented in Table 8.4.

Table 8.4: Coded values of process variables for cold backward extrusion of collapsible tube.

Factors	***Coded levels***				
	–2	***–1***	***0***	***+1***	***+2***
Extrusion force (x_1) KN	380	477.5	575	672.5	770
Stroke length (x_2), mm	341	341.75	342.5	343.25	344
Diameter of slug (x_3), mm	19	19.75	20.50	21.25	22
Back pressure (x_4), MPa	50	68.75	87.5	106.25	125

To study the effect of process parameters *i.e.*, F_e, S_l, d_s, and P_b, on performance measures *i.e.*, power (P_w), deviation in extruded length (L_e), thickness of diaphragm (T_d), and nozzle thickness (T_n) a second-order polynomial response is fitted into the following Eq. (3.29). Equations (8.6) - (8.9) are then derived by determining the values of the coefficients using the least square technique for power (P_w), extruded length (L_e), thickness of diaphragm (T_d), and nozzle thickness (T_n) respectively.

$$\begin{aligned} P_w &= 5166.443 + 279.723x_1 + 212.949x_2 + 175.132x_3 + \\ &254.782x_4 - 56.597x_1^2 - 64.815x_2^2 - 28.354x_3^2 + \\ &44.014x_4^2 - 11.029x_1x_2 - 31.295x_1x_3 + 16.021x_1x_4 + \\ &13.097x_2x_3 + 12.230x_2x_4 + 36.667x_3x_4 \end{aligned} \tag{8.6}$$

$$\begin{aligned} L_e &= 151.375 + 3.648x_1 + 4.126x_2 - 2.2958x_3 - 2.6083x_4 + \\ &0.4927x_1^2 + 0.5302x_2^2 - 0.4985x_3^2 - 1.2797x_4^2 - 0.3175x_1x_2 + \\ &0.0025x_1x_3 - 0.805x_1x_4 - 1.2475x_2x_3 - 1.135x_2x_4 + 0.3325x_3x_4 \end{aligned} \tag{8.7}$$

$$\begin{aligned} T_d &= 0.13605 - 0.013737x_1 - 0.009462x_2 + 0.0035958x_3 - \\ &0.020004x_4 - 0.006703x_1^2 - 0.008028x_2^2 - 0.0072406x_3^2 - \\ &0.0079031x_4^2 - 0.0018937x_1x_2 - 0.000943x_1x_3 - 0.00263125x_1x_4 + \\ &0.0000687x_2x_3 - 0.0006437x_2x_4 - 0.00094375x_3x_4 \end{aligned} \tag{8.8}$$

$$\begin{aligned} T_n &= 1.28395 - 0.0637666x_1 - 0.0216x_2 + 0.000775x_3 - \\ &0.0742333x_4 + 0.0074541x_1^2 - 0.0069958x_2^2 + 0.00041666x_3^2 - \\ &0.01538333x_4^2 + 0.016825x_1x_2 - 0.015x_1x_3 - 0.0483375x_1x_4 + \\ &0.004875x_2x_3 + 0.0088625x_2x_4 - 0.0249625x_3x_4 \end{aligned} \tag{8.9}$$

To test whether the data are well fitted in model or not, the calculated S value of the regression analysis for power, extruded length, thickness of diaphragm, and nozzle thickness are obtained as 118.30, 2.411, 0.011, and 0.054 respectively, which are smaller and R value for these responses are 0.94, 0.88, 0.83, and 0.78 respectively. The R value is moderately high for all responses. Hence, the developed RSM models fit the data well.

Now, to demonstrate the effectiveness of teaching-learning-based optimization algorithm for parameter optimization of cold backward extrusion process following multi-objective optimization model is formulated:

Objectives

***Objective* 1:** The objective is to minimize the extrusion power (P_w) as given by the Eq. (8.6).

***Objective* 2:** Minimization of deviation of extrusion length (d_L) from its specified value

$$d_L = 130 - L. \tag{8.10}$$

Where, L is length of tube as given by Eq. (8.7)

***Constraints*:** The constraints considered in this work are mainly to ensure the dimensional accuracy of the product. The constraint on surface roughness is not considered as the test component considered in work is subjected to post extrusion processes like hardening and painting. Hence, following two constraints are considered.

***a) Thickness of diaphragm*:** For proper functioning of the extruded tube for the intended purpose, the thickness of diaphragm (T_d) should be between 0.10-0.15 mm. This constraint can be expressed as:

$$T_d - 0.10 \geq 0 \tag{8.11}$$

$$0.15 - T_d \geq 0 \tag{8.12}$$

Where, T_d is as given by Eq. (8.8).

***b) Thickness of nozzle*:** For proper functioning of the extruded tube for the intended purpose, thickness of nozzle (T_n) should be between 1.15 to 1.525 mm. This constraint can be expressed as:

$$T_n - 1.15 \geq 0 \tag{8.13}$$

$$1.525 - T_n \geq 0 \tag{8.14}$$

Where, T_n is as given by Eq. (8.9).

The variable bounds for the four variables considered in this work are:

$$380 \leq F_e \leq 770 \text{ (kN)} \tag{8.15}$$

$$341 \leq S_l \leq 344 \text{ (mm)} \tag{8.16}$$

$$19 \leq d_s \leq 22 \text{ (mm)} \tag{8.17}$$

$$50 \le P_b \le 125 \text{ (N/mm}^2\text{)} \tag{8.18}$$

In coded form, the bounds for all variables (x_i) lie between –2 to 2.

The combined objective function considering all constraints is then formulated as given by Eq. (8.19).

$$Min\ Z = w_1 \times \frac{P_w}{P_{w_{min}}} + w_2 \times \frac{d_L}{d_{L_{min}}} + CV_1 \times P_1 + CV_2 \times P_2 + CV_3 \times P_3 + CV_4 \times P_4 \tag{8.19}$$

Where, w_1 and w_2 are the weights assigned to the two objectives. In present example, equal weights (w_1= 0.5 and w_2 = 0.5) are considered. Pw_{min} is the minimum value of extrusion power obtained when single objective optimization problem considering only extrusion power as an objective was solved for the given constraints. dL_{min} is the minimum value of deviation of extrusion length obtained when single objective optimization problem considering only deviation of extrusion length as an objective, was solved for the given constraints.CV_1 to CV_4 are the amount of constraint violations for constraints given by Equations (8.11) to (8.14) and P_1 to P_4 are static penalties determined for constraints CV_1 to CV_4 respectively.

Now the TLBO algorithm is applied to solve this optimization problem. Based on several trial runs, a population size of 10 is adopted for this example.

Step 1: ***Set initial conditions***

Let N = 10.

Step 2: ***Generate initial population (Initial scores of students)***

For 10 students the marks in each subject and scores are as shown in Table 8.5.

Step 3: ***Improve score in teacher phase***

In teacher phase students learn from teacher. A teacher is considered to be the most knowledgeable and hence represented by the global best solution in population. Since the problem is of minimization the best solution (Teacher) in Table 8.6 is solution 1 with score 1.301. Each student will learn from teacher to improve score as given by Equations (8.1) to (8.3). The updated marks and new scores of the students are shown in Table 8.6. It can be seen solutions 2, 7, and 10 are retained as the updated solutions (not shown here) were inferior to their respective original solutions.

Table 8.5: Generation of initial solutions.

Student roll number (solution no.)	*Marks in subject 1 (x_1)*	*Marks in subject 2 (x_2)*	*Marks in subject 3 (x_3)*	*Marks in subject 4 (x_4)*	P_w	d_L	*Score (z)*
1	–0.583	–1.115	0.493	0.193	4790.05	14.5155	1.301
2	–0.125	–0.749	–1.542	0.924	4815.54	15.9222	5.078
3	–0.644	0.963	0.555	–0.115	5190.11	22.1208	1.393
4	0.857	–1.153	–0.511	–0.921	4788.51	22.3328	1.331
5	–1.085	0.52	–0.434	–0.815	4559.48	19.8916	1.310
6	1.361	–1.343	0.744	–0.081	5106.59	22.6603	1.382
7	0.218	–1.309	–0.914	1.915	5237.48	10.2224	28.203
8	–1.173	1.331	1.411	1.216	5647.88	12.9391	16.344
9	0.124	0.425	1.897	–1.787	5059.99	17.0018	1.353
10	–0.494	–1.920	–1.807	0.914	4145.00	9.9740	16.281

Table 8.6: Scores of students after implementing teacher phase.

Student roll number (solution no.)	*Marks in subject 1 (x_1)*	*Marks in subject 2 (x_2)*	*Marks in subject 3 (x_3)*	*Marks in subject 4 (x_4)*	P_w	d_L	*Score (z)*
1	–0.969	–1.515	0.664	0.522	4605.37	12.02	1.263
2	–0.125	–0.749	–1.542	0.924	4815.54	15.92	5.078
3	–0.913	0.676	0.625	0.264	5181.46	18.79	1.379
4	0.436	–1.469	–0.471	–0.359	4683.03	18.81	1.301
5	–1.220	0.011	–0.380	–0.541	4437.29	16.58	1.282
6	0.968	–1.952	0.905	0.006	4828.81	19.65	1.327
7	0.218	–1.309	–0.914	1.915	5237.49	10.22	28.203
8	–1.173	1.331	1.411	1.216	5647.89	12.94	16.335
9	–0.013	–0.201	2.398	–1.198	5024.85	12.80	1.331
10	–0.494	–1.920	–1.807	0.914	4145.00	9.97	16.281

Step 4: *Improve scores in Learner phase*

Updated marks and new scores of the students after implementing Lerner phase are shown in Table 8.7. It may be noted that solution number 2, 5, 8, 9, and 10 obtained in teacher phase are retained as corresponding updated solutions in learner phase are inferior.

At the end of first generation, the teacher (*gBest*) performance is improved from z=1.301 to z”=1.235. Average fitness value is also improved from 7.40 in initial population to 4.68.

Table 8.7: Scores of students after implementing learner phase.

Student roll number (solution no.)	***Marks in subject 1 (x_1)***	***Marks in subject 2 (x_2)***	***Marks in subject 3 (x_3)***	***Marks in subject 4 (x_4)***	P_w	d_L	***Score (z)***
1	–1.292	–1.486	1.133	0.147	4454.34	10.81	1.235
2	–0.125	–0.749	–1.542	0.924	4815.54	15.92	5.078
3	–0.704	0.666	0.898	–1.462	4899.90	19.89	1.339
4	0.521	–1.844	–0.893	–0.792	4414.60	17.68	1.255
5	–1.220	–0.457	–0.380	–0.541	4437.29	16.58	1.254
6	–0.043	–0.926	–0.375	–0.264	4694.60	17.62	1.315
7	0.579	–0.831	–0.500	0.499	5141.19	19.78	1.376
8	–1.173	1.331	1.411	1.216	5647.89	12.94	16.344
9	–0.013	–0.201	2.398	–1.198	5024.85	12.80	1.331
10	–0.494	–1.920	–1.807	0.914	4145.00	9.97	16.281

Step 5: *Repeat steps 2 to 4 till termination criteria is met*

The steps 2 to 4 are repeated till a termination criterion is met.

The convergence graph is shown in Fig. 8.3.

As can be seen the algorithm converge to the optimum solution in 10 generations only and the optimum solution is:

- Extrusion force (F_e) = 380 KN (Level: –2)
- Stroke length (S_l) = 341 mm (Level: –2)
- Slug diameter (d_s) = 19.71 mm (Level: –1.053)
- Back pressure (P_b) = 54.97 MPa (Level: –1.734)

Values of performance measures corresponding to this optimum solution are:

- Extrusion pressure (P_w) = 3252.33 KN
- Deviation of tube length (L_d) = 2.44 mm
- Thickness of diaphragm (t_d) = 0.1
- Thickness of nozzle (t_n) = 1.402
- Combined objective function (z) = 1.0157

To evaluate the robustness of the algorithm, 30 independent runs of the algorithm are conducted. For these 30 runs, the mean of the objective function values is 1.0158, best objective function value is 1.0154 and the standard deviation of objective function values is 0.0006 only.

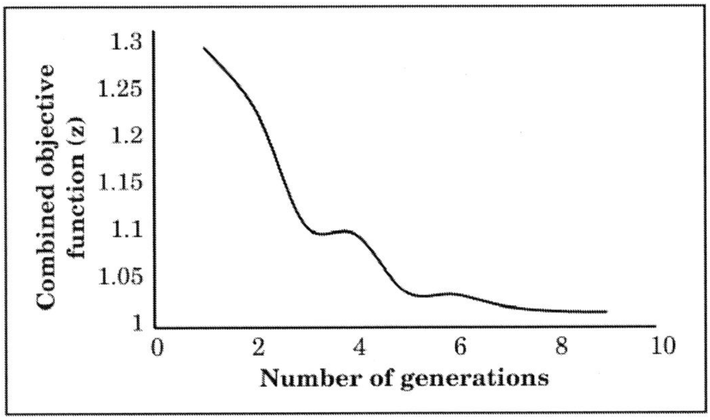

Fig. 8.3: Convergence graph of TLBO algorithm.

8.4. Optimization of Supply Chain System in Multi-Product and Multi-Supplier Scenario

Supply chain management (SCM) is the management of the flow of goods and services and includes all processes that transform raw materials into final products. It involves the active streamlining of a business's supply-side activities to maximize customer value and gain a competitive advantage in the marketplace. SCM represents an effort by suppliers to develop and implement supply chains that are as efficient and economical as possible. Supply chains cover everything from production to product development to the information systems needed to direct these undertakings. It is well known that supply chain management is an integral part of most businesses and is essential to company success with respect to the following:

a) *Boost customer service*: by ensuring supply of right product in right quantity of right quality at right place and right time to the customer.

b) *Reduce operating costs*: by decreasing purchase cost, production cost and total supply chain cost.

c) *Improve financial position of organization*: by increase profit leverage, decreasing cost of the fixed assets, and increasing the cash flow.

The operational issues of a supply chain include determination of sales quantity, transportation cost, quantity discount, production capacity, storage capacity, starting-operation quantity, total service level, total quality level, and the system costs including purchasing cost, holding cost, and cost due to shortages. To assure that customer demand can be satisfied without delays, it is necessary to keep inventory in the supply chain, which helps to absorb the impact of random events. A stream of research has emerged in recent years aiming at the coordination of inventory replenishment decisions among individual companies to benefit the entire supply network by reducing the total cost of the system. The integrated supply chain models used to coordinate the supply chain system with multiple buyers and multiple vendors is usually a multi-objective mixed integer non-linear programming problems. Due to the complex nature of this problem, researchers are employing various meta-heuristics to solve this problem. In this section, the application of TLBO algorithm is demonstrated to solve supply chain optimization problem considering following two different types of models.

8.4.1. *Model 1: Multi-objective mixed integer non-linear programming (MOMINLP) model without shortage*

In this model the main assumption is that the shortages are not allowed. The three objectives considered in this model are minimization of the total cost of the supply chain, maximization of total quality level of supply chain, and maximization of service level of supply chain as discussed below (Rezaei and Davoodi, 2011):

Objective 1: *Minimization of cost of supply chain (Z_1)*

The cost of the supply chain is sum of purchase cost, ordering cost, inventory holding cost, and transportation cost as given by Eq. (8.20).

$$Min\, Z_1 = \sum_i \sum_j \sum_t p_{ij} x_{ijt} + \sum_j \sum_t o_j e^{-\beta_j \sum_{k=1}^{t} yjk} y_{jt} +$$

$$\sum_{i}\sum_{t} h_i\left(\sum_{k=1}^{t}\sum_{j} x_{ijk} - \sum_{j}\left(1 - s_{ij0}e^{\gamma_{ij}t}\right)x_{ijt} - \sum_{k=1}^{t} d_{ik}\right)$$

$$+\sum_{j}\sum_{t} g_j\left[\frac{\sum_i w_i x_{ijt}}{v_j}\right] \tag{8.20}$$

Where, p = net purchase cost, o = ordering cost, β = ordering cost reduction rate, y = 1 (if order is placed), y = 0 (otherwise), h = holding cost, s = service level, g = service level growth rate, d = demand, g = transportation cost per vehicle, w = space occupied by product, v = vehicle capacity.

Objective 2: *Maximization of total quality level (Z_2)*

The overall quality level of all the products ordered from all suppliers in all periods should be maximized. This model also takes into account the growth trend of quality level for each product offered by each supplier as an exponential time dependent function (l) as given by Eq. (8.21).

$$Max\, Z_2 = \sum_{i}\sum_{j}\sum_{t} f_{ij0}e^{\lambda_{ijt}} x_{ijt} \tag{8.21}$$

Where, f = product quality level, l = quality level growth rate

Objective 3: *Maximization of total service level (Z_3)*

The overall service level of all the products ordered from all suppliers in all periods should be maximized. This model also takes into account the growth trend of service level for each product offered by each supplier as an exponential time dependent function (g) as given by Eq. (8.22).

$$Max\, Z_3 = \sum_{i}\sum_{j}\sum_{t} s_{ij0}e^{\gamma_{ijt}} x_{ijt} \tag{8.22}$$

The above objectives are subjected to following five constraints:

1) Demand: The ordered quantity should be sufficient to meet demand in order to prevent shortages. This constraint is given by Eq. (8.23) below.

$$\sum_{k=1}^{t}\sum_{j} x_{ijk} - \sum_{k=1}^{t} d_{ik} \geq \sum_{j}\left(1 - s_{ij0}e^{\gamma_{ij}t}\right)x_{ijt} \tag{8.23}$$

***2) Charging ordering cost*:** This constraint is to ensure that the buyer can't place order without ordering cost. This constraint is given by Eq. (8.24) below.

$$\left(\sum_{k=1}^{T} d_{ik}\right)y_{jt} - x_{ijt} \geq 0 \tag{8.24}$$

***3) End of planning inventory*:** This constraint is to ensure that the inventory level at the end should become zero. This constraint is given by Eq. (8.25) below.

$$\left(\sum_{k=1}^{T}\sum_{j} x_{ijk} - \sum_{j}\left(1 - s_{ij0}e^{\gamma_{ij}T}\right)x_{ijt} - \sum_{k=1}^{T} d_{ik}\right) = 0 \tag{8.25}$$

***4) Storage capacity*:** This constraint is to ensure that the space occupied by total number of products should not exceed the storage capacity of a supplier. This constraint is given by Eq. (8.26) below.

$$\sum_{i} w_i\left(\sum_{k=1}^{t}\sum_{j} x_{ijk} - \sum_{j}\left(1 - s_{ij0}e^{\gamma_{ij}t}\right)x_{ijt} - \sum_{k=1}^{t} d_{ik}\right) \leq W \tag{8.26}$$

where, W = *total storage capacity*

***5) Supplier capacity*:** This constraint is to ensure that the number of products ordered to a supplier should not exceed his capacity. This constraint is given by Eq. (8.27) below.

$$x_{ijt} \leq c_{ij} \tag{8.27}$$

Where, c_{ij} is the supplier capacity.

The variables considered in this work are the quantities ordered for each product from each supplier in specified period (x_{ijt}).

8.4.2. *Model 2: MOMINLP model with backorder*

In this model the assumption is that shortages and backordering is allowed. Hence, considering the backordering cost, the objective function for total cost of system (Z_1) can be written as:

$$Min\, Z_1 = \sum_i \sum_j \sum_t p_{ij} x_{ijt} + \sum_j \sum_t o_j e^{-\beta_j \sum_{k=1} yjk} y_{jt} +$$

$$\sum_i \sum_t h_i \left(\sum_{k=1}^{t} \sum_j x_{ijk} - \sum_j \left(1 - s_{ij0} e^{\gamma_{ij} t}\right) x_{ijt} - \sum_{k=1}^{t} d_{ik} \right)^+ \qquad (8.28)$$

$$\sum_i \sum_t b_i \left(\sum_{k=1}^{t} \sum_j x_{ijk} - \sum_j \left(1 - s_{ij0} e^{\gamma_{ij} t}\right) x_{ijt} - \sum_{k=1}^{t} d_{ik} \right)^-$$

$$\sum_j \sum_t g_j \left\lceil \frac{\sum_i w_i x_{ijt}}{v_j} \right\rceil$$

Where, b_i is the backordering cost of product *i*.

Remaining two objective functions *i.e.*, maximization of total quality level of supply chain (Z_2), and maximization of service level of supply chain (Z_3) are same as considered in section 8.4.1 and given by equations (8.21) and (8.22) respectively. This model is subjected to the constraints of charging ordering cost, end of planning inventory, storage capacity, and supplier capacity as given by Equations (8.24) to (8.27) respectively.

The variables considered in this model are also the quantities ordered for each product from each supplier in specified period ().

Now, to demonstrate the effectiveness of proposed TLBO algorithm, an application example is considered for a supply chain with 3 products, and 5 suppliers with 4 replenishment periods. The product and supplier data is given in Tables 8.8 to 8.10.

As TLBO algorithm is an algorithm-specific parameter-less algorithm, only population size and number of generations need to be specified to run the algorithm. Based on several trial runs, the population size decided for the present example is 10 and the numbers of generations are set to 115.

Table 8.8: Demand for three products in four periods (Rezaei and Davoodi, 2011).

	Product	***Period (t)*** 1	2	3	4
d_{it}	1	454	540	675	755
	2	327	320	290	285
	3	645	650	637	663

Table 8.9: Holding cost, backordering cost, and occupied space of each product (Rezaei and Davoodi, 2011).

Product (i)	h_i	b_i	w_i
1	12	17	0.75
2	35	38	0.85
3	7	10	0.60

Table 8.10: Other data related to products and/or suppliers (Rezaei and Davoodi, 2011).

	Product (i)	***Supplier (j)*** 1	2	3	4	5
p_{ij}	1	66	73	62	60	71
	2	121	134	127	142	122
	3	28	26	29	25	23
f_{ij0}	1	0.95	0.98	0.92	0.96	0.99
	2	0.89	0.92	0.95	0.90	0.91
	3	0.93	0.87	0..88	0.83	0.87
S_{ij0}	1	0.95	0.97	0.94	0.99	0.98
	2	0.92	0.99	0.91	0.99	0.92
	3	0.99	0.98	0.99	0.97	0.99
c_{ij}	1	550	480	420	890	790
	2	570	85	450	235	525
	3	825	430	850	390	720
l_{ij}	1	0.001	–0.002	0.0015	–0.001	–0.001
	2	0	0.0015	–0.001	0.002	0.002
	3	–0.001	0.002	0.0015	0.0015	0.0012
g_{ij}	1	0.0013	0.0011	–0.001	0.001	–0.0011
	2	0	0.0014	0.002	0.0015	0.002
	3	–0.0011	0.0016	–0.001	–0.002	0.0011
$\hat{a}_j$		0.08	0.1	0.09	0.12	0.09
o_j		45000	64500	37250	53400	42250
v_j		85	150	85	100	150
g_j		33500	55200	33500	38400	55200

$W = 150$

8.4.3. *Results of optimization for supply chain model without allowing shortage*

For supply chain model discussed in section 8.4.1, Rezaei and Davoodi (2011) obtained set of Pareto optimal solution and declared the three solutions out of those as best solutions with respect to three objectives each. For comparison purpose, this problem is solved as single objective optimization problem considering each objective separately. The optimum ordered quantities ordered for each product from each supplier in specified period (x_{ijt}) when minimization of the total cost is considered as a sole objective are presented in Table 8.11. The objective function values for the above optimum combinations using TLBO along with those obtained using GA are presented in Table 8.12. As shown in Table 8.12, it is observed that the results obtained by TLBO are superior to GA with respect to all three objective functions when minimization of cost is considered as an objective.

Table 8.11: The optimum ordered quantities when minimization of cost is considered as a sole objective.

x_{ijk}	*Value*	x_{ijk}	*Value*
x_{111}	65	x_{233}	112
x_{112}	140	x_{234}	104
x_{113}	169	x_{241}	0
x_{114}	176	x_{242}	0
x_{121}	31	x_{243}	19
x_{122}	123	x_{244}	10
x_{123}	152	x_{251}	174
x_{124}	175	x_{252}	126
x_{131}	22	x_{253}	117
x_{132}	95	x_{254}	109
x_{133}	140	x_{311}	198
x_{134}	162	x_{312}	205
x_{141}	370	x_{313}	203
x_{142}	146	x_{314}	201
x_{143}	176	x_{321}	107
x_{144}	198	x_{322}	107
x_{151}	58	x_{323}	104
x_{152}	146	x_{324}	112
x_{153}	175	x_{331}	120
x_{154}	197	x_{332}	119
x_{211}	107	x_{333}	116
x_{212}	176	x_{334}	124
x_{213}	134	x_{341}	97

Table 8.11: *Contd...*

x_{ijk}	*Value*	x_{ijk}	*Value*
x_{214}	153	x_{342}	97
x_{221}	50	x_{343}	94
x_{222}	12	x_{344}	102
x_{223}	12	x_{351}	180
x_{224}	12	x_{352}	179
x_{231}	113	x_{353}	177
x_{232}	121	x_{354}	184

Table 8.12: Comparison of objective function values using TLBO and GA for minimization of cost.

Method	Z_1 *(to be minimize)*	Z_2 *(To be maximized)*	Z_3 *(To be maximized)*
GA (Rezaei and Davoodi, 2011)	3013904	5786.101	6076.555
TLBO	3007996	6822.87	7145.18

The optimum ordered quantities ordered for each product from each supplier in specified period (x_{ijt}) when maximization of the total quality level is considered as a sole objective are presented in Table 8.13. The objective function values for the above optimum combinations using TLBO along with those obtained using GA are presented in Table 8.14. As shown in Table 8.15, it is observed that the results obtained by TLBO are superior to GA with respect to all three objective functions when maximization of quality level is considered as an objective.

The optimum ordered quantities ordered for each product from each supplier in specified period (x_{ijt}) when maximization of the total service level is considered as a sole objective are presented in Table 8.15. The objective function values for the above optimum combinations using TLBO along with those obtained using GA are presented in Table 8.16. As shown in Table 8.16, it is observed that the results obtained by TLBO are superior to GA with respect to all three objective functions when maximization of service level is considered as an objective.

It is also observed that all specified constraints are well satisfied by the results obtained by using TLBO. It can be seen from Tables 8.11, 8.13, and 8.15, that the quantity ordered to each supplier is less than its supply capacity (c_{ij}) mentioned in Table 8.10. The results also show that the quantity ordered for each product in given period is exactly same as that of its demand for that period (d_{it}) mentioned in Table 8.8.

It clearly indicates that there is no shortage. Moreover, as the supply and demand are exactly matching, it is just in time strategy making the inventory holding cost and floor space requirement theoretically zero.

Table 8.13: The optimum ordered quantities when maximization of quality level is considered as a sole objective.

x_{ijk}	*Value*	x_{ijk}	*Value*
x_{111}	46	x_{233}	112
x_{112}	157	x_{234}	99
x_{113}	170	x_{241}	87
x_{114}	177	x_{242}	59
x_{121}	5	x_{243}	52
x_{122}	143	x_{244}	38
x_{123}	156	x_{251}	121
x_{124}	175	x_{252}	121
x_{131}	0	x_{253}	113
x_{132}	117	x_{254}	100
x_{133}	142	x_{311}	177
x_{134}	161	x_{312}	177
x_{141}	349	x_{313}	177
x_{142}	165	x_{314}	177
x_{143}	179	x_{321}	107
x_{144}	197	x_{322}	108
x_{151}	146	x_{323}	108
x_{152}	165	x_{324}	108
x_{153}	179	x_{331}	177
x_{154}	197	x_{332}	177
x_{211}	90	x_{333}	177
x_{212}	120	x_{334}	177
x_{213}	113	x_{341}	97
x_{214}	144	x_{342}	98
x_{221}	27	x_{343}	98
x_{222}	27	x_{344}	98
x_{223}	20	x_{351}	176
x_{224}	6	x_{352}	176
x_{231}	120	x_{353}	176
x_{232}	119	x_{354}	176

Table 8.14: Comparison of objective function values using TLBO and GA for maximization of quality level.

Method	Z_1 *(to be minimize)*	Z_2 *(To be maximized)*	Z_3 *(To be maximized)*
GA[70]	3228229	5847.146	6113.339
TLBO	3191895	7055.78	7390.25

Table 8.15: The optimum ordered quantities when maximization of service level is considered as a sole objective.

x_{ijk}	*Value*	x_{ijk}	*Value*
x_{111}	48	x_{233}	112
x_{112}	156	x_{234}	99
x_{113}	170	x_{241}	87
x_{114}	176	x_{242}	59
x_{121}	7	x_{243}	52
x_{122}	142	x_{244}	38
x_{123}	156	x_{251}	120
x_{124}	175	x_{252}	120
x_{131}	0	x_{253}	113
x_{132}	117	x_{254}	100
x_{133}	142	x_{311}	177
x_{134}	162	x_{312}	177
x_{141}	351	x_{313}	177
x_{142}	164	x_{314}	177
x_{143}	178	x_{321}	107
x_{144}	197	x_{322}	108
x_{151}	141	x_{323}	108
x_{152}	164	x_{324}	108
x_{153}	178	x_{331}	177
x_{154}	197	x_{332}	177
x_{211}	90	x_{333}	177
x_{212}	120	x_{334}	177
x_{213}	113	x_{341}	97
x_{214}	144	x_{342}	98
x_{221}	27	x_{343}	98
x_{222}	27	x_{344}	98
x_{223}	20	x_{351}	177
x_{224}	7	x_{352}	177
x_{231}	120	x_{353}	177
x_{232}	119	x_{354}	177

Table 8.16: Comparison of objective function values using TLBO and GA for maximization of service level.

Method	Z_1 *(to be minimize)*	Z_2 *(To be maximized)*	Z_3 *(To be maximized)*
GA (Rezaei and Davoodi, 2011)	3235358	5777.225	6120.463
TLBO	3134568	7054.79	7408.5

8.4.4. *Results of multi-objective optimization for supply chain model with backorder*

For supply chain model discussed in section 8.4.2, Rezaei and Davoodi (2011) obtained set of Pareto optimal solution and declared the three solutions out of those as best solutions with respect to three objectives each. This model is solved using TLBO by converting a multi-objective optimization problem into a single objective optimization problem by formulating the normalized combined objective function (*Z*) as given by Eq. (8.29).

Minmize $Z = (w_1 \times Z_1 / Z_{1min}) - (w_2 \times Z_2 / Z_{2max}) - (w_3 \times Z_3 / Z_{3max})$ (8.29)

Where, Z_{1min} = Minimum value of total cost of supply chain obtained when single objective optimization problem considering only total cost of supply chain as an objective, was solved for the given constraints = 2738338

Z_{2max} = Maximum value of total quality level obtained when single objective optimization problem considering only total quality level as an objective was solved for the given constraints = 6867.23

Z_{3max} = Maximum value of total service level obtained when single objective optimization problem considering only total service level as an objective was solved for the given constraints = 7175.09

w_1, w_2, and w_3 are the weightages assigned to the objective functions Z_1, Z_2, and Z_3 respectively. In the present example, equal weightages for all objectives are assumed. The results of optimization using TLBO are presented in Table 8.17.

The combined objective function values for the above optimum combinations using TLBO and GA are presented in Table 8.17. It may be noted that for GA, the combined objective function value is calculated considering the individual best values of three objective functions obtained by Rezaei and Davoodi (2011). Hence, no solution in set of Pareto optimal solutions provided by Rezaei and Davoodi (2011) can produce better result than presented in Table 8.18. It is observed that the results obtained by TLBO are superior to GA with respect to two objectives, Z_2 and Z_3 and slightly inferior with respect to the objective Z_1. However, the combined objective function (*Z*) obtained by using TLBO is much better as compared to that obtained by using GA.

Table 8.17: Results of optimization for minimization of combined objective function.

x_{ijk}	***Value***	x_{ijk}	***Value***
x_{111}	0	x_{233}	76
x_{112}	25	x_{234}	79
x_{113}	88	x_{241}	0
x_{114}	164	x_{242}	0
x_{121}	0	x_{243}	37
x_{122}	70	x_{244}	0
x_{123}	1	x_{251}	203
x_{124}	35	x_{252}	150
x_{131}	69	x_{253}	55
x_{132}	126	x_{254}	191
x_{133}	6	x_{311}	190
x_{134}	17	x_{312}	283
x_{141}	252	x_{313}	336
x_{142}	58	x_{314}	155
x_{143}	360	x_{321}	0
x_{144}	500	x_{322}	0
x_{151}	226	x_{323}	9
x_{152}	371	x_{324}	14
x_{153}	355	x_{331}	327
x_{154}	192	x_{332}	348
x_{211}	101	x_{333}	233
x_{212}	221	x_{334}	163
x_{213}	186	x_{341}	0
x_{214}	116	x_{342}	0
x_{221}	0	x_{343}	4
x_{222}	0	x_{344}	20
x_{223}	40	x_{351}	181
x_{224}	0	x_{352}	260
x_{231}	141	x_{353}	68
x_{232}	64	x_{354}	233

Table 8.18: Comparison of objective function values using TLBO and GA

Method	Z_1	Z_2	Z_3	Z
GA	2738338	5873.361	6143.507	–0.700
TLBO	3015769	6867.23	7175.09	–0.898

Table 8.19 provides the optimum values of various elements for model 1 and model 2. As shown in Table, as in model 2, the constraint on demand satisfaction for each period is reduced; the ordering cost is reduced. However, this results in increase in other cost elements *i.e.,*

purchase cost, holding cost, backorder cost, and transportation cost. As shown in Table 8.19, the floor space constraint is well satisfied in case of both models.

Table 8.19: Optimum values of various elements obtained for model 1 and model 2.

Element	*Purchase cost*	*Ordering cost*	*Holding cost*	*Transport-ation cost*	*Back order cost*	*Floor space*
Model 1	471314	657286.6	0	1879000	0	0
Model 2	472191	628088.4	1240	1912500	1751	103

It is observed that the results obtained by proposed TLBO algorithm are superior to those obtained using GA for both the models. This may be due to the fact that the algorithm-specific parameter-less TLBO uses the best solution of the iteration to modify the existing solution as in PSO and it does not divide the population as in case of ABC. The proposed algorithm may be easily applied to various optimization problems of the industrial environment such as determination of optimal ordering and pricing policies; supplier selection and order lot sizing; locating distribution centres and allocating customers demands in supply chains; vehicle-routing problems in transportation, job shop scheduling, flow shop scheduling, design of cellular manufacturing systems, design of facility location networks; inventory control; production planning and control; etc. In general, the proposed algorithm may be easily customized to suit the optimization of any system involving large number of variables and objectives.

8.5. Variants of Teaching Learning Based Optimization Algorithm

A number of methods have been proposed to improve the robustness and computational efficiency of TLBO algorithm. Few of them are discussed below.

8.5.1. *Nonlinear inertia weighted teaching-learning-based optimization*

Wu *et al.* (2015) proposed a novel improved TLBO, which is called nonlinear inertia weighted TLBO. A nonlinear inertia weighted factor is introduced into the basic TLBO to control the memory rate of learners, and another dynamic inertia weighted factor is used to replace the original random number in teacher phase and learner phase. So, as a result, this algorithm may have faster convergence speed and higher calculation accuracy for most of these optimization problems than the basic TLBO.

The basic TLBO algorithm is based on teaching-learning phenomenon of a classroom. In the teacher phase, the teacher tries to shift the mean of the learners towards himself or herself by teaching. In the learner phase, learners improve their knowledge by interaction among themselves. In the process of the teaching-learning, learners improve their level by accumulating knowledge. In other words, they learn new knowledge based on existing knowledge. In the real world, the teacher tends to wish that his or her students should achieve the knowledge equal to him in fast possible time. But it is impossible for a student because of his or her forgetting characteristics. In fact, a student usually forgets a part of existing knowledge due to the physiological phenomena of the brain. With increasing the iteration numbers of learning, more and more existing knowledge will be remembered. The sharpest increase occurs after the first try and then gradually evens out, meaning that less and less new knowledge is retained after each repetition. Like the forgetting curve, the learning curve is exponential. So it is necessary to add a memory weight to the existing knowledge of the student for simulating this learning scenario. According to this phenomenon, a nonlinear inertia weighted factor is introduced into Equations (8.3) to (8.5) in the basic TLBO, and this factor is considered as memory weighted factor which controls the memory rate of learners. This nonlinear inertia weighted factor will scale the existing knowledge of the learner for computing the new value. In contrast to the basic TLBO, in this algorithm the part of previous value of the learner is decided by a weighted factor while computing the new learner value.

Accordingly, to meet the characteristic of memory to conform to the learning curve, the nonlinear inertia weighted factor (*i.e.,* memory rate) is nonlinearly increased from $w_{\min}$ to 1.0 over time, whose value is given as:

$$w = 1 - e^{\left(\frac{-iter^{2}}{2\times\left(\frac{maxiter}{8}\right)^{2}}\right)} . (1 - w_{min}) \tag{8.30}$$

Here '*iter*' is the current iteration number, '*maxiter*' is the maximum number of allowable iterations, and w_{min} is the minimum value of nonlinear inertia weighted factor. The value w_{min} should be above 0.5, or the individuals are worse due to remembering too little existing knowledge at first. Hence, if the value w_{min} is too small, the algorithm could not converge to the true global optimal solution. The nonlinear inertia weighted factor w is applied to the new equations shown by Eq. (8.32) and Eq. (8.33). In this modified TLBO, the individuals try to sample diverse zones of the search space during the early stages of the search. During the later stages, the individuals adjust the movements of trial solutions finely so that they can explore the interior of a relative small space.

In the teacher phase, in order to obtain a new set of better learners, the difference between the existing mean result and the corresponding result of the teacher is added to the existing population of learners. Similarly, to obtain a new set of better learners in the learner phase, two learners are selected randomly, and the difference between their result of each corresponding subject is added to the existing learner. As Equations (8.3) to (8.5) shown, the difference value added to the existing learner is formed from the difference of result and the random number. Therefore, in the teacher and learner phases, the difference value is decided by the random number to a large extent. In this method, the random number is modified as follows:

$$r' = 0.5 + \frac{r}{2} \tag{8.31}$$

where r is a uniformly distributed random number within the range Eq. (8.31) generates a random number in the range [0.5, 1] .r' is called as dynamic inertia weighted factor. Therefore, the mean value of the random number r is raised from 0.5 to 0.75. This increases the probability of stochastic variations and enlarges the difference value added to the existing learners, so as to improve population diversity, avoid prematurity in the search process, and increase the ability of the basic TLBO to escape from local optima. On the multimodal function surface, the original random weighed factor leads to most of the populations clustering near a local optimum point. However, the population with new dynamic inertia weight has more chances to jump out of the local optima and continuously move towards the global optimum point until a true global optimum is reached.

With the nonlinear inertia weighted factor and the dynamic inertia weighted factor, the new set of improved learners can be expressed by using equation in the teacher phase as:

$$Difference\_Mean_{j,k,i} = r_i' \, (X_{j,kbest,i} - T_F M_{j,i}) \qquad (8.32)$$
$$X'_{j,k,i} = w.\, X_{j,k,i} + Difference\_Mean_{j,k}$$

and the new set of improved learners can be expressed by using equation in the learner phase are:

$$X''_{j,P,i} = w.X'_{j,P,i} + r_i' \, (X'_{j,P,i} - X'_{j,Q,i}), \text{ If } X'_{total-P,i} < X'_{total-Q,i} \qquad (8.33)$$

$$X''_{j,P,i} = w.X'_{j,P,i} + r_i' \, (X'_{j,Q,i} - X'_{j,P,i}), \text{ If } X'_{total-Q,I} < X'_{total-P,i}$$

Where, w and r_i' are as given by Equations (8.30) and (8.31) respectively.

8.5.2. *Improved teaching-learning-based optimization with the social character of PSO*

The main idea of TLBO is to imitate the teaching-learning process in a class. The teacher tries to disseminate knowledge among the learners to increase the knowledge level of the whole class, and the learners also study knowledge from the others to improve its grade. Algorithm displays that all individuals update their positions based on the distance between one or two times of the mean solution and the teacher in teacher phase. An individual also renewed its position based on the distance between it and a randomly selected individual from the class. Eq. (8.3) indicates that the teacher only improves the grades of the students by using the mean grade of the class; the distance between the teacher and the students is not considered in teacher phase. In PSO, the difference between the current best individual and the individual can help the individual improve its performance. Based on this idea, Feng *et al.* (2016) introduced the method of PSO into TLBO to improve the learning efficiency of the TLBO algorithm. The main change for TLBO is represented in updating equation. Equation (8.34) in TLBO is modified as:

$$X'_{j,k,i} = X_{j,k,i} + r_1(X_{j,kbest,i} - T_F M_{j,i}) + r_2\,(X_{j,kbest,i} - X_{j,k,i}) \qquad (8.34)$$

Where, r_1 and r_2 are the random number in the range. With this modification, the performance of TLBO algorithm might be improved.

In original TLBO algorithm, all genes of each individual should be compared with those of all other individuals for removing the duplicate individual. This operator is a heavy computation cost process, and the function evaluations required are not clearly known. In our opinion, duplication testing is not needed in every generation especially in the beginning of evolution. Assume that any two individuals have the same genes in generation, and the new position of these two individuals might be different when T_F is a random number (1 or 2). At the beginning of evolution, all individuals generate better positions easily. Random operator for may benefit for maintaining diversity of class, but, in the anaphase of evolution, the positions of individuals might be close to each other. When the mean solution equals the best solution, the individual does not change ($T_F = 1$) or the large change of genes might destroy the individual in large degree ($T_F = 2$) so that the better individual is difficult to be generated. To decrease the computational effort of comparing all the individuals, the removing of the duplicate individual process in the original TLBO is deleted in the improved TLBO algorithm and a mutation operator according to the best fitness of successive generations is introduced. If the best fitness of continuous generations is not changed or changed slightly, an individual will be randomly selected according to mutation possibility to be mutated. To maintain the global performance of the algorithm, the best individual is not mutated.

Steps of Improved Teaching-Learning-Based Optimization with the Social Character of PSO are discussed below (Feng *et al.,* 2016):

Step 1: Generate initial population

Step 2: Evaluate the individual, select the best individual as the teacher, and calculate the mean solution of the population.

Step 3: For each individual update its position using Eq. (8.34). Old solution is replaced by new one if better.

Step 4: For each individual, randomly select another individual, update its position according to Eq. (8.4) and Eq. (8.5). Old solution is replaced by new one if better.

Step 5: Execute mutation operator for the population.

Step 6: If the end condition of TLBO-PSO is not satisfied, the algorithm will go back to Step 2, or it is terminated.

9

Fuzzy Logic Applications in Optimization

9.1. Introduction

Computational intelligence techniques such as artificial neural networks, fuzzy logic, and evolutionary algorithms are popular research subjects, since they can deal with complex engineering problems which are difficult to solve by classical methods.

Fuzzy Logic (FL) is a method of reasoning that resembles human reasoning. The approach of fuzzy logic imitates the way of decision making in humans that involves all intermediate possibilities between digital values YES and NO. The conventional logic block that a computer can understand takes precise input and produces a definite output as TRUE or FALSE, which is equivalent to human's YES or NO. The inventor of fuzzy logic, Zadeh (1965), observed that unlike computers, the human decision making includes a range of possibilities between YES and NO, such as CERTAINLY YES, POSSIBLY YES, CANNOT SAY, POSSIBLY NO, CERTAINLY NO etc.

The fuzzy logic works on the levels of possibilities of input to achieve the definite output.

Fuzzy logic system has four main parts:

- *Fuzzification module* – It transforms the system inputs, which are crisp numbers, into fuzzy sets. It splits the input signal into five steps such as –

- *Knowledge base* – It stores IF-THEN rules provided by experts.
- *Inference engine* – It simulates the human reasoning process by making fuzzy inference on the inputs and IF-THEN rules.
- *Defuzzification module* – It transforms the fuzzy set obtained by the inference engine into a crisp value.

Figure 9.1 shows diagrammatic representation of fuzzy logic system.

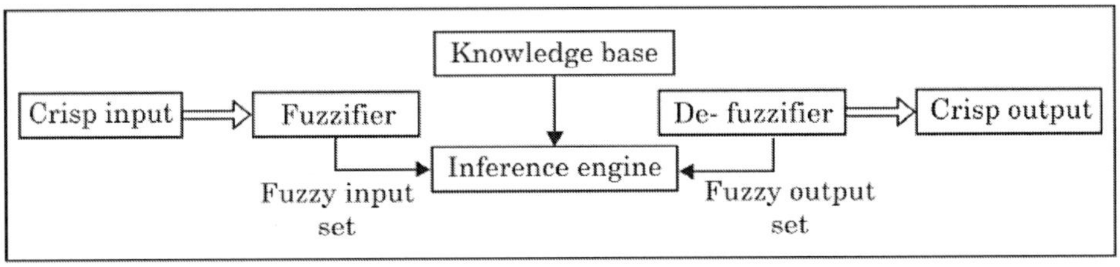

Fig. 9.1: Fuzzy logic schematic (Boumella *et al.,* 2012).

A fuzzy set is distinct from a crisp set in that it allows its elements to have a degree of membership. The core of a fuzzy set is its membership function: a surface or line that defines the relationship between a value in the set's domain and its degree of membership. In particular, according to the original idea of Zadeh (1965), membership of an element x to a fuzzy set A, denoted as $\mu A(x)$ or simply $A(x)$, can vary from 0 (full non-membership) to 1 (full membership), *i.e.,* it can assume all values in the interval [0, 1]. Clearly, a fuzzy set is a generalization of the concept of a set whose membership function takes on only two values {0,1}. The fuzziness in fuzzy logic is best characterized by its membership function which represents the degree of truth in fuzzy logic. Selection of membership function largely depends on the data pattern. Commonly used membership functions are:

(1) *Triangular membership function*: The triangular membership function shown in Fig. 9.2 for A: X- [0, 1], is given by three parameters according to:

$$\mathrm{A}(X,a,b,c)=\begin{cases} 0, & x<a \\ (x-a)/(b-a), & a\le x\le b \\ (c-x)/(c-b), & b\le x\le c \\ 0, & x>c \end{cases}$$

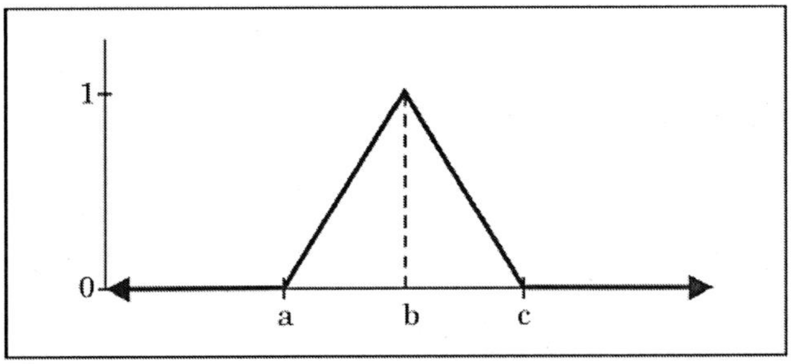

Fig. 9.2: Triangular membership function

(2) *Trapezoidal function*: The trapezoidal membership function shown in Fig. 9.3 for π: X – [0, 1] is given by four parameters according to,

$$\prod(x:a,b,c,\delta)=\begin{cases}0, & x<a\\(x-a)/(b-a), & a\leq x\leq b\\1, & b\leq x\leq c\\(\delta-x)/(\delta-c), & c<x\leq\delta\\0, & x>\delta\end{cases}$$

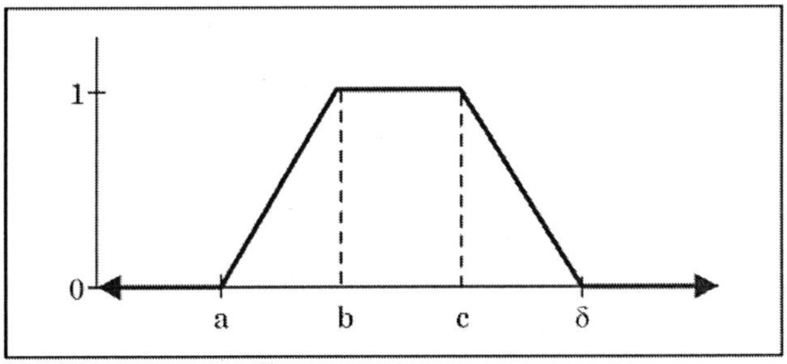

Fig. 9.3: Trapezoidal membership function

(3) *Gaussian function*: The membership definition for a Gaussian function shown in Fig. 9.4 for G: - [0, 1] is given by two parameters as:

$$G(x:a,b)=e^{-b(x-a)^2}$$

Where a is the midpoint and b reflects the slope value. Note that b must be positive and that the function never reaches zero. The Gaussian function can also be extended to have different left and right slops. We then have three parameters in

$$G(x: a, b_l, b_r) = \begin{cases} e^{-b_l(x-a)^2}, & x \leq a \\ e^{-b_r(x-a)^2}, & x > a \end{cases}$$

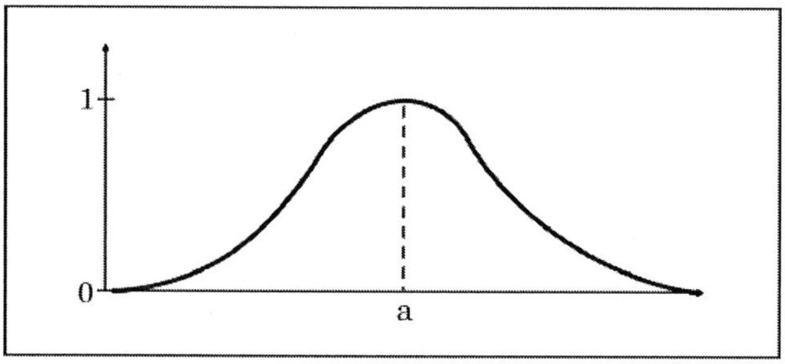

Fig. 9.4: Gaussian membership function

The concept of fuzzy logic is demonstrated below through an example to estimate the loan eligibility of a salaried person. Assuming that the loan eligibility of an individual depends on two factors: salary and age, fuzzy logic can be applied to estimate loan eligibility. The input factors are (i) salary of a person and (ii) age of a person and the response is loan eligibility. Let the membership function for both input factors be triangular with 3 levels: [Low, Moderate, High] for salary of a person and [Young, Adult, Old] for age of a person and membership function for response is also triangular but with 5 levels: [Very poor, Poor, Average, High, Very High].

The triangular membership function *(m)* for age of a person is shown in Fig. 9.5 and that for salary is shown in Fig. 9.6 below.

The triangular membership function *(m)* for loan eligibility at five levels is shown in Fig. 9.7.

The fuzzy rules can be defined as:

- *If individual is OLD and salary is LOW then loan eligibility is VERY POOR*

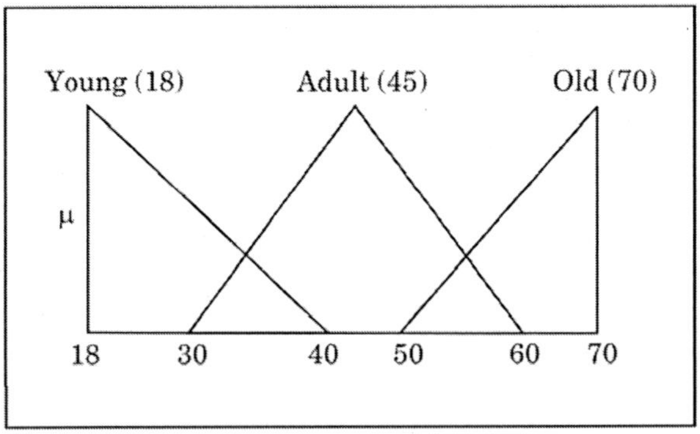

Fig. 9.5: Membership function for age

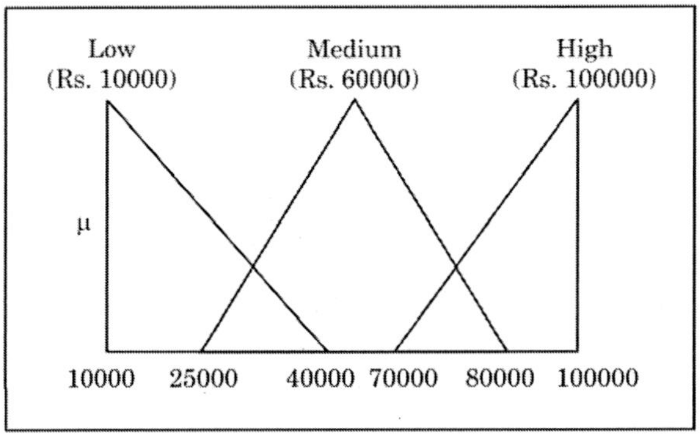

Fig. 9.6: Membership function for monthly salary

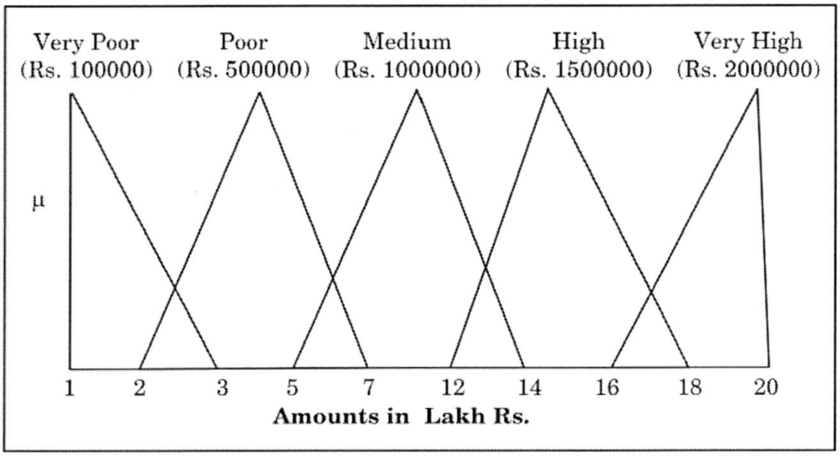

Fig. 9.7: Membership function for loan eligibility

- *If individual is OLD and salary is MODERATE then loan eligibility is MEDIUM*
- *If individual is OLD and salary is HIGH then loan eligibility is MEDIUM*
- *If individual is ADULT and salary is LOW then loan eligibility is POOR*
- *If individual is ADULT and salary is MODERATE then loan eligibility is HIGH*
- *If individual is ADULT and salary is HIGH then loan eligibility is HIGH*
- *If individual is YOUNG and salary is LOW then loan eligibility is HIGH*
- *If individual is YOUNG and salary is MODERATE then loan eligibility is VERY HIGH*
- *If individual is YOUNG and salary is HIGH then loan eligibility is VERY HIGH*

Now, consider an individual having age 53 years and monthly salary of Rs. 35000. We have to estimate the loan amount for which he/she is eligible. It can be seen from Fig. 9.8 that, age of 53 years has a membership value of 0.467 for 'Adult' and membership value of 0.150 for 'Old'.

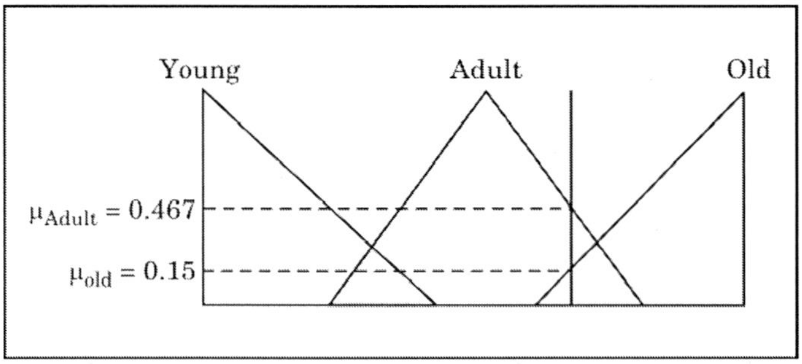

Fig. 9.8: Membership values for age = 53 years

Similarly, monthly salary of Rs. 35000 has membership values of 0.166 for 'Low' and 0.285 for 'Moderate'. Thus, the memberships of 'Age' for the three levels are (Young = 0, Adult = 0.467, and Old = 0.15) and those for 'Monthly salary' are: (Low = 0.166, Moderate = 0.285, and High = 0). The membership values for the response *i.e.,* loan eligibility are then determined based on fuzzy rules and fuzzy set operations. The fuzzy set operations are:

***Union*:** The membership function of the Union of two fuzzy sets A and B with membership functions μ_A and μ_B respectively is defined as the maximum of the two individual membership functions. This is called the maximum criterion.

$$\mu_{AUB} = \text{Max}\ (\mu_A, \mu_B)$$

The Union operation in Fuzzy set theory is the equivalent of the OR operation in Boolean algebra.

***Intersection*:** The membership function of the Intersection of two fuzzy sets A and B with membership functions μ_A and μ_B respectively isdefined as the minimum of the two individual membership functions. This is called the *minimum* criterion.

$$\mu_{A \cap B} = \text{Min}\ (\mu_A, \mu_B)$$

The Intersection operation in Fuzzy set theory is the equivalent of the AND operation in Boolean algebra.

***Complement*:** The membership function of the Complement of a Fuzzy set A with membership function μ_A is defined as the negation of the specified membership function. This is called the *negation* criterion.

$$\mu_{\bar{A}} = 1 - \mu_A$$

The Complement operation in Fuzzy set theory is the equivalent of the NOT operation in Boolean algebra.

As the AND operation (mentioned in fuzzy rules) is represented by Intersection in fuzzy sets and is performed by considering the minimum of μ_A and μ_B, the resultant membership values for loan eligibility are: (Very Poor = 0.15, Poor = 0.166, Medium = 0.15, and High= 0.285, Very High: 0). The crisp value of loan eligibility is then obtained by de-fuzzification. Various commonly used de-fuzzification methods are:

***Center of Sums Method*:** Center of sums is the fastest de-fuzzification method. This process involves the algebraic sum of individual output fuzzy sets. The only drawback of this method is that intersecting areas are added twice.

Center of gravity Method: This method provides a crisp value based on the center of gravity of the fuzzy set. The total area of the membership function distribution used to represent the combined control action is divided into a number of sub-areas. The area and the center of gravity or centroid of each sub-area is calculated and then the summation of all these sub-areas is taken to find the de-fuzzified value for a discrete fuzzy set.

Center of Area / Bisector of Area Method: This method calculates the position under the curve where the areas on both sides are equal. The BOA generates the action that partitions the area into two regions with the same area.

Weighted Average Method: This method is valid for fuzzy sets with symmetrical output membership functions and produces results very close to the COA method. This method is less computationally intensive. Each membership function is weighted by its maximum membership value.

Maxima Method: This method considers values with maximum membership. There are different maxima methods with different conflict resolution strategies for multiple maxima such as: First of Maxima Method, Last of Maxima Method, Mean of Maxima Method.

The center of gravity (or centroid) method is demonstrated here for de-fuzzification. The area of each fuzzy set is obtained from its corresponding membership value. For example, the area for set 'High' is the area of hatch section as shown in Fig. 9.9.

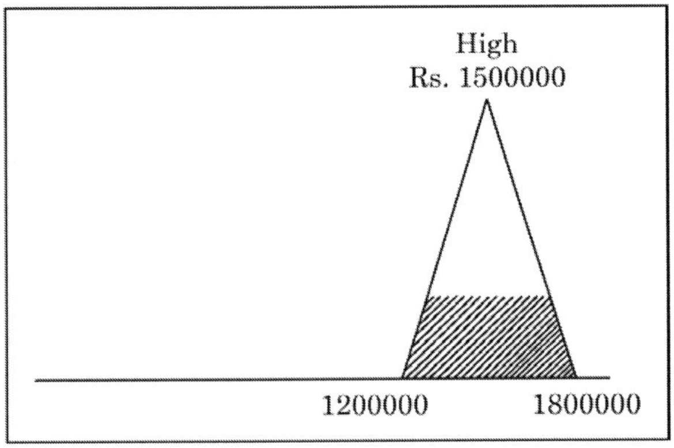

Fig. 9.9: Area for fuzzy set 'High'

In similar manner, the contribution (areas) of other sets can be obtained. Those values are 41625, 83720, 124900, 146600, and 0 respectively for the levels Very Poor, Poor, Medium, High, and Very High respectively. The total area is thus 396845 units. The crisp score of loan eligibility value then can be obtained as:

$$Loan\ Eligibility\ Value = \frac{41625 \times 100000 + 83720 \times 500000 + 124900 \times 1000000 + 146600 \times 1500000 + 0 \times 2000000}{41625 + 83720 + 124900 + 146600 + 0} = 984824$$

Thus for a person having age of 53 years and monthly salary of Rs. 35000 is eligible to get a loan amount of Rs. 984824.

9.2. Collaboration of Fuzzy Logic and Evolutionary Algorithms

There are three possible ways for integrating fuzzy logic and evolutionary algorithms. The first one involves the application of evolutionary algorithms for solving optimization and search problems related with fuzzy systems, called as evolutionary fuzzy systems. The second one concerns the use of fuzzy tools and fuzzy logic-based techniques for modelling different evolutionary algorithm components and adapting evolutionary algorithm control parameters, with the goal of improving performance. The evolutionary algorithms resulting from this integration are called fuzzy evolutionary algorithms. The third one is to determine the weights of objective functions in priori approach of multi-objective optimization. These three approaches of collaborations of fuzzy logic and evolutionary algorithms are discussed below.

9.2.1. *Evolutionary fuzzy systems*

Although, fuzzy logic has been described as a practical, robust, economical, and intelligent alternative for modeling and control of complex systems, the success of the fuzzy logic can only be realized if quality expert knowledge exists. In conventional applications of fuzzy logic, there is not yet a systematic way of acquiring such expert knowledge. Quality expert knowledge may not exist, or even if it exist, it is not clear whether the expert would be able to objectively present his knowledge in terms of a constrained set of rules and membership functions, and whether such expertise would indeed be optimal. In solving this paradox, fuzzy logic has been complemented by various strategies such as neural networks, fuzzy clustering, gradient-based methods and evolutionary optimization algorithms. Among the above strategies, evolutionary optimization algorithms have been particularly appealing to various the scientific circles, primarily because such algorithms allow autonomous adaptation/optimization of fuzzy systems without human intervention, are not easily trapped in locally optimal solutions, allow parallel exploitation of an optimization parameter space, and do not require gradient evaluation of the objective function. Evolutionary fuzzy systems are hybrid fuzzy systems where evolutionary optimization algorithms are used to optimize/adapt fuzzy expert knowledge.

The evolutionary fuzzy system approaches can be divided into two processes, tuning and learning. If there exists a knowledge base, evolutionary algorithms can be applied to tune the fuzzy rule base system parameters for improving the fuzzy rule base system performance but without changing the existing rule base. Various approaches for tuning the fuzzy rule base system are discussed below (Herrera and Lozano, 2009).

***a) Tuning of knowledge Base parameters*:** The tuning process carried out considering the whole knowledge base obtained (the preliminary database and the derived rule base), is used a posterior to adjust the membership function parameters. Nevertheless, the tuning process only adjusts the shapes of the membership functions and not the number of linguistic terms in each fuzzy partition, which remains fixed from the beginning of the design process.

***b) Adaptive inference systems*:** The main aim of this approach is the use of parameterized expressions in the Inference System, sometimes called Adaptive Inference Systems, for getting higher cooperation among the fuzzy rules and therefore more accurate fuzzy models without loosing the linguistic rule interpretability.

***c) Adaptive defuzzification methods*:** The most popular technique in practice, due to its good performance, efficiency and easier implementation, is to apply the defuzzification function to every inferred rule fuzzy set (getting a characteristic value) and to compute them by a weighted average operator. This method introduces the possibility of using parameter based average functions, and the use of evolutionary algorithms can allow us to adapt the defuzzification methods.

The type of evolutionary fuzzy systems is to learn Knowledge base components. Various approaches for learning of fuzzy rule base system are discussed below.

***a) Rule Base learning*:** Existing database is used to formulate initial rule base. The performance of this rule base (fitness function) is evaluated and evolutionary algorithms can be applied to improve the fitness value and thereby defining the most effective rile base for fuzzy system.

***b) Rule Base selection*:** A very large rule base and an excessive number of rules makes it difficult to understand the fuzzy rule base system behavior. A fuzzy rule set may consist of: irrelevant rules, redundant rules, erroneous rules and conflictive rules, which perturb the fuzzy rule base system performance when they co-exist with others. To face this problem we can use a genetic rule selection process for obtaining an optimized subset of rules from a previous fuzzy rule set, by selecting some of them. The rule base selection can be combined with tuning approaches, to try to get a good rule set together with a tuned set of parameters.

***c) Simultaneous learning of rule base and data base*:** This approach provides learning of the two components of the knowledge base *i.e.,* database and rule base simultaneously. This approach can result into possibility of generating better definitions of fuzzy systems but there is a need to deal with a larger search space that makes the learning process more difficult and slow.

9.2.2. *Fuzzy evolutionary algorithms*

To overcome the drawbacks of traditional optimization methods, researchers are employing evolutionary optimization methods such as genetic algorithm (GA), particle swarm optimization (PSO), harmony search algorithm (HS), artificial immune algorithm, memetic algorithm (MA) etc. to solve complex models of manufacturing systems. However, the major difficulty in application of these algorithms lies in their selection of appropriate algorithm-specific parameters such as crossover probability, mutation probability, scaling function, selection function, etc. in case of GA; inertia coefficient and acceleration coefficients in case of PSO; and harmony memory consideration rate and pitch adjusting rate in case of HS algorithm.

The performance of all the above mentioned algorithms is greatly influenced by their respective algorithm-specific parameters in addition to the common control parameters such as population size and number of generations. Selection of suitable values of these algorithm-specific parameters for a particular application is itself is a complex optimization problem. Fuzzy logic-based techniques can be used to tune these parameters in a self-adapting manner to improve the performance of evolutionary algorithms.

To demonstrate this concept, consider an optimization of following model using particle swarm optimization (PSO) algorithm:

$$Minimize\ z = x_1^2 - 3x_2 + 10 \text{ such that } 0 \leq x_{1,}x_2 \leq 3.$$

As discussed earlier, PSO has three algorithm specific parameters: Inertia weight (w), Cognitive parameter (C_1) and Social parameter (C_2). However, since C_1 and C_2 are coupled with the random numbers which changes continuously in every generation, do not require further adjustment (as it will result into another random number only). Hence, inertia coefficient (w) is considered as the only parameter which needs adjustment to improve the performance of PSO algorithm. The performance of PSO algorithm can be evaluated in each step through performance measures such as 'Increment in global best value (*IGB*)' and 'Deviation of the fitness values in the population (*DFV*)'. Now, the fuzzy logic can be used to determine the best value of inertia coefficient (w) in a particular generation based on the values of *IGB* and *DFV* obtained in that generation. Let the membership function for both *IGB* and *DFV* be triangular with 3 levels: [LOW, MEDIUM, HIGH] and the membership function for inertia coefficient also to be triangular and with 5 levels: [VERY LOW, LOW, MEDIUM, HIGH, VERY HIGH]. The Increment in global best value of 4 is considered LOW, that of 8 is considered MEDIUM and value of 12 is considered as HIGH. The deviation of the fitness values of 4 is considered as LOW, that of 12 is considered as MEDIUM and value of 20 is considered as HIGH. The triangular membership function (m) for IGB is shown in Fig. 9.10 and that for DFV is shown in Fig. 9.11 below.

The inertia coefficient (w) is considered as VERY LOW at value 0.00, LOW at value 0.28, MEDIUM at value 0.56, HIGH at value 0.72, and VERY HIGH at value 1.00. The triangular membership function (m) at five levels for inertia coefficient (w) is shown in Fig. 9.12.

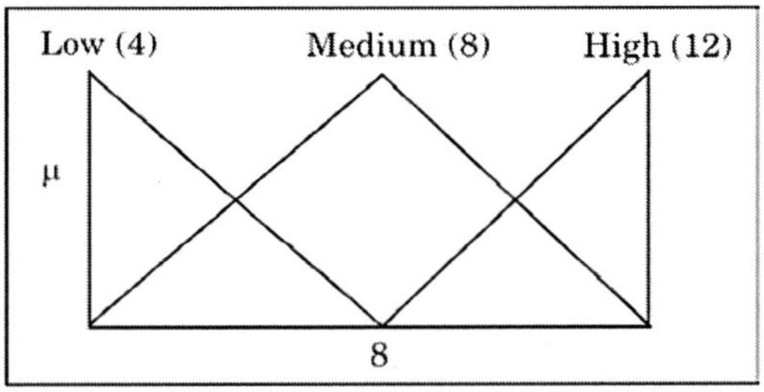

Fig. 9.10: Membership function for Increment in global best value (*IGB*).

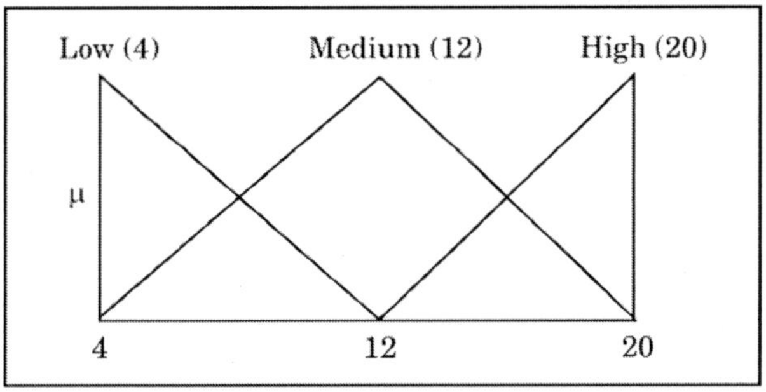

Fig. 9.11: Membership function for deviation of the fitness values (DFV) in the population

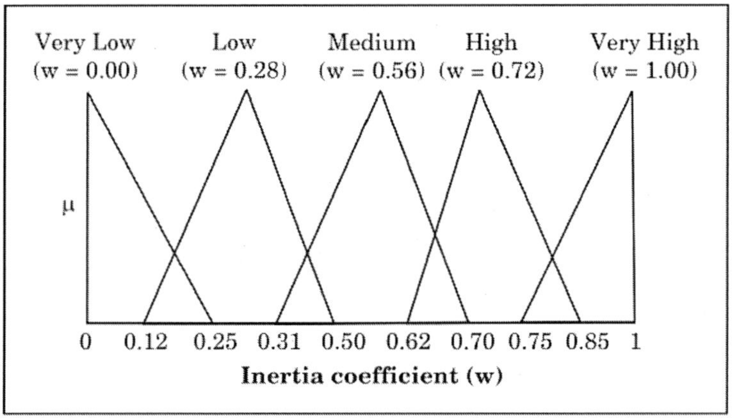

Fig. 9.12: Membership function for inertia coefficient (w).

The fuzzy rules can be defined as (Marco et al., 2018)

1. If IGB is Low and DFV Low then w is Medium
2. If IGB is Low and DFV Medium then w is Low

3. If IGB is Low and DFV High then w is Very High
4. If IGB is Medium and DFV Low then w is Very Low
5. If IGB is Medium and DFV Medium then w is Medium
6. If IGB is Medium and DFV High then w is High
7. If IGB is High and DFV Low then w is Low If IGB is High and DFV Medium then w is High
8. If IGB is High and DFV High then w is Medium

Considering five birds in a swarm, the Table 9.1 shows initial position of birds (x_1, x_2) and their fitness values (f_i). The global best solution (*gBest*) is solution 1. The updated positions are shown in Table 9.2.

Table 9.1: Initial population of solutions

Solution no.	*1 (gBest)*	*2*	*3*	*4*	*5*
x_1	2	2.7	1.9	2.5	3
x_2	1.8	2	0.8	1	2.2
f_i	8.6	11.29	11.21	13.25	12.4

Table 9.2: Updated positions of the birds

Solution no.	*1*	*2*	*3*	*4*	*5 (gBest)*
x_1	2.28	2.934	2.077	2	1.78
x_2	1.872	1.115	1.21	1.386	2.643
f_i	9.582	15.27	10.68	9.861	5.24

In the set of updated solutions, solution 5 is global best (*gBest*). The *IGB* is then computed as:

$$IGB = gBest\ f_{k-1} - gBest\ f_k \tag{9.1}$$

Where, $gBest\ f_k$ is the fitness value of global best solution at i^{th} generation. Thus, for the present case, *IGB* = 8.6-5.24 = 3.36.

The deviation of the fitness values (*DFV*) is computed as:

$$DFV = \sqrt{\frac{1}{N}\sum_{i=1}^{N}\left(\left|f_i - \overline{f}\right|\right)^2} \tag{9.2}$$

Where, N is population size, f_i is the fitness value of ith solution in the current population, and $\bar{f}$ is the average fitness value of the current population. Thus for the updated population, DFV = 5.101.

Now, IGB = 3.36 has membership values of 0.44 for LOW and 0.56 for MEDIUM as shown in Fig. 9.13.

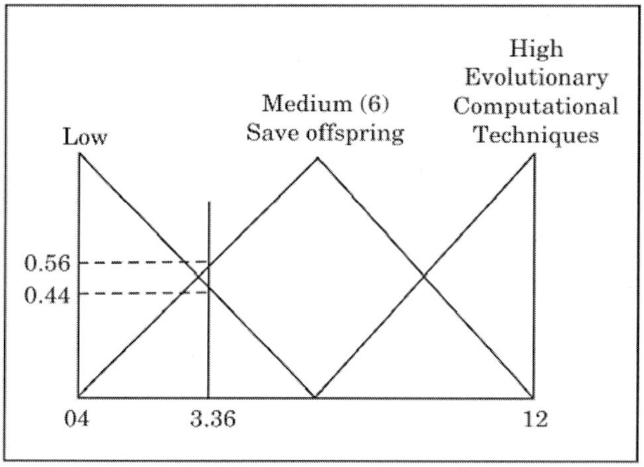

Fig. 9.13: Membership values for IGB = 3.36.

Similarly, DFV of 5.101 has membership values of 0.725 for 'MEDIUM' and 0.275 for 'HIGH' as shown in Fig. 9.14

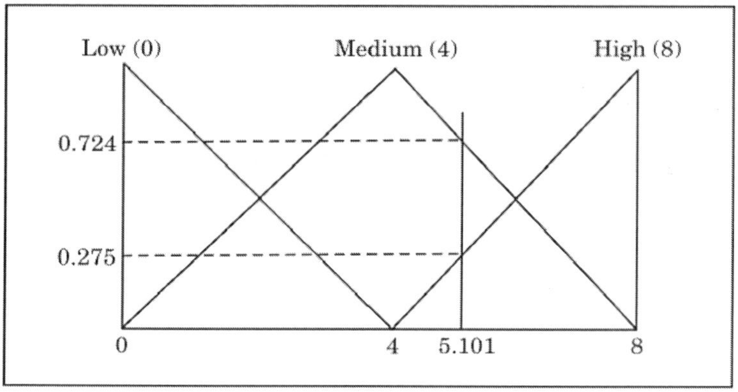

Fig. 9.14: Membership values for DFV = 5.101.

Thus, the memberships of '*IGB*' for the three levels are (LOW = 0.44, MEDIUM = 0.56, and HIGH = 0) and those for '*DFV*' are: (LOW = 0, MEDIUM = 0.724, and HIGH = 0.275). The membership values for the response *i.e.,* inertia coefficient (w) are then determined based on fuzzy rules and fuzzy set operations.

As the AND operation used in the given Rule Base is represented by Intersection in fuzzy sets and is performed by considering the minimum of μ_A and μ_B, the resultant membership values for 'w' are: (VERY LOW = 0, LOW = 0.44, Medium = 0.56, and High = 0.275, Very High: 0.275). The crisp value of loan eligibility is then obtained by de-fuzzification. The de-fuzzification is carried out using center of gravity (or centroid) method. The area of each fuzzy set is obtained from its corresponding membership value. For example, the area for set 'MEDIUM' is the area of hatch section as shown in Fig. 9.15.

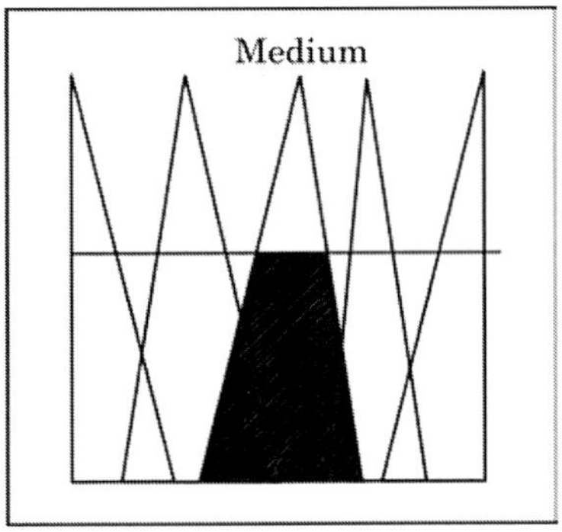

Fig. 9.15: Area for fuzzy set MEDIUM.

In similar manner, the contribution (areas) of other sets can be obtained. Those values are 0, 0.130, 0.157, 0.054, and 0.059 respectively for the levels Very Poor, Poor, Medium, High, and Very High respectively. The total area is thus 0.4 units. The crisp score of inertia coefficient (w) then can be obtained as:

$$w = \frac{0 \times 0 + 0.130 \times 0.28 + 0.157 \times 0.56 + 0.054 \times 0.72 + 0.059 \times 1}{0 + 0.130 + 0.157 + 0.054 + 0.059} = 0.555$$

Thus the value of inertia coefficient is updated to a value of 0.555 over the assumed value of 0.65 for the first generation. Now, keeping the values of C_1 and C_2 same as that of first generation and value of w

as 0.555, the velocity and the position of the every bird in the swarm is updated in the next generation. It can be seen that the fuzzy PSO algorithm offers better performance in terms of solution accuracy and variance of the final results especially for hard noisy problems.

9.2.3. *Fuzzy rating scale for qualitative weights in multi-objective optimization*

There are several methods for multi-objective optimization such as:

- ***Non-preference methods*:** These methods do not assume any information about the importance of objectives, but an heuristic is used to find a single optimum solution. They do not make any attempt to find multiple Pareto-optimal solutions.
- ***Posteriori methods*:** Posteriori methods use preference information of each objective and iteratively generate a set of Pareto-optimal solutions.
- ***Interactive methods*:** These methods use the preference information progressively during the optimization process.
- ***Priori methods*:** Priori methods use more information about the preferences of objectives and they usually find one preferred Pareto-optimal solution.

These methods of multi-objective optimization are discussed in detail in chapter 10. This section presents the application of fuzzy logic to determine the weights of the objectives in case of priori methods. In priori methods, the weights of the objective functions are decided by decision maker/process planner before the optimization process. The general formulation of this optimization model using priori approach is:

$$Z = \sum_{i=1}^{n} w_i \times \frac{Z_i}{Z_i^*} \tag{9.3}$$

Where, Z is value of combined objective function, n is the number of objectives, Z_i = value of ith objective function, Z_i^* is the threshold value of ith objective function. Z_i may be positive or negative depending on the nature of objective function (minimization or maximization).

In Eq. (9.3), the weights (w_i) should be quantitative. However, usually, it is difficult for decision maker to provide the crisp value of these weights, instead he/she has qualitative judgement about the expected behavior of each objective function. For example surface roughness should be relatively low. To convert the weights expressed qualitatively into quantitative form, usually Likert scale, originally introduced by Likert (1932), is used widely. A well designed Likert scale should state the opinion, attitude, or belief being measured in clear terms and use the appropriate wording for scale points. However, Likert scale suffers from following serious drawbacks:

- Likert scales fail to approximate intervals of ordinal data
- The respondents are forced to make a choice from the given options that may not match their exact responses. They have to either select an answer from an insufficient range of responses or respond to an "acceptable" answer in the closed format. This miss-matching further worsens the information distortion problem.
- Significant amount of information is lost and/or distorted due to the built-in limitations of the Likert method.

To overcome the above mentioned limitations of the Likert scale, researchers have proposed a fuzzy scale. Fuzzy scale is a flexible method to obtain a numerical value from ordinal variables. Fuzzy Scale design could easily incorporate the knowledge of an expert in the fuzzification of input and in the building-up of control block-rules, as they allow for the best adaptability. Many popular scales, indexes, and measures may be conceived as fuzzy sets denoting graded concepts. The combination of computational intelligence with Fuzzy Scale could represent a useful route for the analysis of individual behavior or judgments. The procedure for developing the fuzzy scale is discussed below.

9.3. Development of Fuzzy Scale

Chen and Hwang (1992) proposed an approach to solve multi-attribute decision making (MADM) problems in a fuzzy environment. The approach is of two steps. In the first step, fuzzy data is converted into crisp scores. Then, as the next step, this data in the form of decision matrix is used to rank the alternatives by using MADM methods. In the following section, this method has been explained. This method logically converts linguistic terms into their corresponding fuzzy numbers.

Fuzzy numbers 'W' are converted into the crisp score as:

$$\mu_{\max}(x) = \begin{Bmatrix} x, 0 \le x \le 1 \\ 0, otherwise \end{Bmatrix}$$

$$\mu_{\min}(x) = \begin{Bmatrix} 1-x, 0 \le x \le 1 \\ 0, otherwise \end{Bmatrix} \tag{9.4}$$

The maximum and minimum fuzzy numbers should be selected in a manner that they can be fitted automatically into comparison scale. The left score of the fuzzy number is calculated as below:

$$\mu_L(W_i) = Sup_x \left\lfloor \mu_{\min}(x)^{W_i(x)} \right\rfloor \tag{9.5}$$

Where Sup_x is the supremum of fuzzy set. The term supremum should not be confused with the term maximum. The maximum is the largest member of the set, while the supremum is the smallest upper bound of the set. Consider a set A = $\{x \mid x < 5\}$. Here, the maximum is not well defined but supremum is well defined by value of 5 which represents the smallest upper bound of this set.

By using the formula given by Eq. 9.5, the left score is converted into crisp number between 0 and 1. It is the maximum value of the intersection of fuzzy number W_j and the minimum fuzzy number. Similarly, the right score can be converted into crisp number by using the following formula.

$$\mu_R(W_i) = Sup_x \left\lfloor \mu_{\max}(x)^{W_i(x)} \right\rfloor \tag{9.6}$$

The total score is calculates as:

$$\mu_T(W_i) = \frac{[\mu_R(W_i) + 1 - \mu_L(W_i)]}{2} \tag{9.7}$$

To demonstrate the procedure for development of fuzzy scale and to compare its performance with Likert scale, an application example of industrial robot selection is discussed here. The goal is to select a robot (out of three robots) performing best with respect to five criteria:

purchase cost (PC), Load carrying capacity (LC), Repeatability error (R), Man Machine Interface (MI), and Programming Flexibility (PF). The objective and subjective information of the attributes is given in Table 9.3. Man–machine interface (MI) and programming flexibility (PF) are expressed subjectively in linguistic terms, and these attributes are assigned objective values using fuzzy numbers provided in Table 9.4. LC, MI, and PF are beneficial attributes (higher values are desirable). *PC* and *R* are non-beneficial attributes (lower values are desirable).

Table 9.3: Robot selection attributes information (Bhangale *et al.,* 2004).

Robot		***PC ($1,000)***	***LC (kg)***	***R (mm)***	***MI***	***PF***
Robot	1	73	48	0.15	A	H
Robot	2	71	46	0.18	AA	VH
Robot	3	75	51	0.14	BA	H

PC: Purchasing cost; LC: Load carrying capacity; R: Repeatability error; MI: Manmachine interface; PF: Programming flexibility; A: Average; AA: Above average; BA: Below average; H: High; VH: Very high.

To convert the qualitative data for 'MI' and 'PF' into quantitative data, 5 point fuzzy scale can be prepared considering qualitative measures as 'Below Average' (BA), 'Average' (A), 'Above Average' (AA), 'High' (H), and 'Very High' (VH). Fuzzy numbers corresponding to these five qualitative measures and the equations to calculate membership function are provided in Table 9.4. The supremum of this fuzzy set can be obtained from Fig. 9.16. The left scores, right scores, and the total scores (Crisp scores) are computed by using Equations (9.5), (9.6), and (9.7) respectively and are shown in Table 9.5. The decision maker can now make use of this 5 point scale to convert qualitative information into quantitative data.

The crisp scores assigned to various linguistic terms in Table 9.5 are based on the conversion scale developed for this given application as shown in Fig. 9.16.

Once the crisp score are assigned to the linguistic variables using conversion scale as shown in Fig. 9.16, the ranking can be done by any of the multiple attribute decision making method such as TOPSIS, Weighted Product method (WPM), Simple additive weights method (SAW), VIKOR, etc, For demonstration purpose simple additive method

Table 9.4: Robot Membership function to calculate Fuzzy numbers.

Qualitative measure intangible factors	*Fuzzy numbers*	*Membership function (μ) of*
Below average	(0, 0, 0.25)	$\mu_w(x) = \begin{cases} 1, x = 0 \\ (0.25 - x)/0.25, 0 \le x \le 0.3 \end{cases}$
Average	(0.1, 0.28, 0.45)	$\mu_w(x) = \begin{cases} (x - 0.1)/0.18, 0.1 \le x \le 0.28 \\ (0.45 - x)/0.17, 0 \le x \le 0.3 \end{cases}$
Above average	(0.35, 0.56, 0.65)	$\mu_w(x) = \begin{cases} (x - 0.35)/0.21, 0.35 \le x \le 0.56 \\ (0.65 - x)/0.09, 0.56 \le x \le 0.65 \end{cases}$
High	(0.5, 0.72, 0.85)	$\mu_w(x) = \begin{cases} (x - 0.5)/0.22, 0.5 \le x \le 0.72 \\ (0.85 - x)/0.13, 0.72 \le x \le 0.85 \end{cases}$
Very high	(0.7, 1, 1)	$\mu_w(x) = \begin{cases} (x - 0.7)/0.3, 0.7 \le x \le 1 \\ 1, x = 1 \end{cases}$

Table 9.5: Crisp value is calculated using right score (mR) and left score (mL).

Qualitative measure of intangible factors	*Right Score (μR)*	*Left score (μL)*	*Crisp score*
Below average	0.2	1	0.1
Average	0.385	0.762	0.311
Above average	0.596	0.537	0.529
High	0.752	0.409	0.671
Very high	1	0.2	0.9

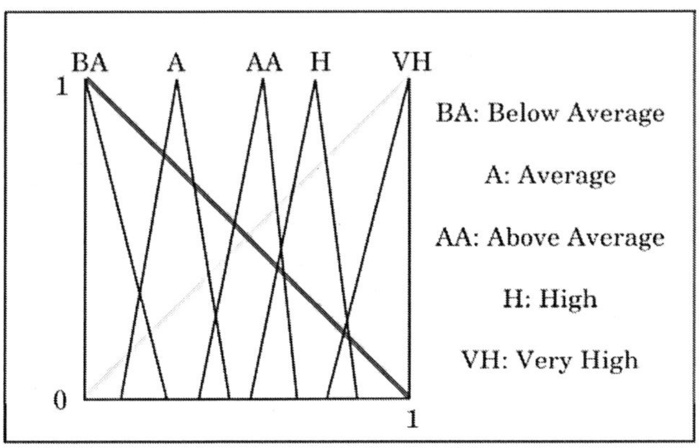

Fig. 9.16: Fuzzy conversion scale for given application.

is used for the present example considering equal weights of all attributes. This is a two-step process. In the first step the attribute data is normalized as shown in Table 9.6. In second step, for each robot the normalized values of attributes are added to get the robot selection index (RSI) as shown in Table 9.7.

Table 9.6: Robot selection attributes using fuzzy scale.

Robot	*PC ($1,000)*	*LC (kg)*	*R (mm)*	*MI*	*PF*
Robot 1	73	48	0.15	0.311	0.671
Robot 2	71	46	0.18	0.529	0.9
Robot 3	75	51	0.14	0.1	0.671

Table 9.7: Robot selection Index (RSI) values using fuzzy scale.

Robot	*PC ($1,000)*	*LC (kg)*	*R (mm)*	*MI*	*PF*	*RSI*
Robot 1	0.972	0.941	0.933	0.587	0.745	4.178
Robot 2	1	0.901	0.777	1	1	4.678
Robot 3	0.946	1	1	0.189	0.745	3.88

RSI: Robot Selection Index;

It is observed from the Table 9.7, that the Robot 2 is best choice with respect to the given attributes as indicated by highest value of RSI.

Now, if the Likert scale having conversion scale BA=1, A=2, AA=3, H=4, VH=5 is used, the robot selection attributes in quantitative form are as shown in Table 9.8.

Table 9.8: Robot selection attributes using Likert scale.

Robot	*PC ($1,000)*	*LC (kg)*	*R (mm)*	*MI*	*PF*
Robot 1	73	48	0.15	2	4
Robot 2	71	46	0.18	3	5
Robot 3	75	51	0.14	1	4

The normalized data and RSI then would be as shown in Table 9.9.

Table 9.9: Robot selection Index (RSI) values using Likert scale

Robot	*PC ($1,000)*	*LC (kg)*	*R (mm)*	*MI*	*PF*	*RSI*
Robot 1	0.972	0.941	0.933	0.666	0.8	4.312
Robot 2	1	0.901	0.777	1	1	4.678
Robot 3	0.946	1	1	0.333	0.8	4.079

SI: Selection Index;

It is observed from the Table 9.9, that the Robot 2 is best choice with respect to the given attributes as indicated by highest value of RSI.

The result (RSI) obtained using fuzzy scale and Likert Scale reveals that there is no significant effect on ranking of the robot and the ranking remains same for analysis using fuzzy scale and Likert Scale. However, it can be seen from Table 9.7 and Table 9.9 that the values of RSI for Robot 1 and Robot 3 are different by employing these two scales. In fact, RSI value represents fitness of the alternative for given criteria. Ranking and fitness are two distinct aspects. Assume that, in present application it is decided that any robot having RSI greater than 4 can be purchased. Then all three robots will be recommended if we use Likert scale but robot 3 will not be recommended if we use fuzzy scale. When such an analysis is essential, fuzzy rating scale will be a much better option as it is based on general perception which is converted into crisp sores as against specific perception in case of Likert scale. This is due to the fact that fuzzy rating method able to capture near exact information without loss of much data.

Chapter 10

Multi-Objective Optimization Methods

10.1. Introduction

The need of optimization in case of manufacturing systems mainly arises due to the various reasons such as to increase the production rate, to increase material removal rate, to improve product quality, to reduce production cost, to reduce tool wear, to reduce operating and maintenance cost etc. Most of these optimization problems are multi-objective with objectives conflicting in nature. In case of multi-objective optimization, instead of obtaining a unique optimal solution, a set of equally good (non-dominating) optimal solutions is usually obtained (Pareto sets). As discusses in Chapter 9, there are mainly two approaches to deal with multi-objective optimization:

i) The priori approach, in which combined objective function is formulated by taking the algebraic sum of weighted normalized values of all objective functions. This process will convert the multi-objective problem into a single objective problem which then can be solved by any suitable method of single objective optimization. The process of optimization using priori approach is presented graphically in Fig. 10.1.

The limitations of the priori approach are

1. The multiple objectives are combined to form a single composite function and the optimum value of this composite function is then obtained, thus in this approach the multi-objective optimization problem is transformed into single objective

optimization problem. Thus the priori approach is not a multi-objective optimization technique in its true sense.

2. The weights are assigned to the individual objective functions before optimum solution is obtained, these weights generally correspond to the priority assigned to each objective, and thus the decision maker should have a comprehensive knowledge about the order of importance of each objective before solving the optimization problem.
3. These algorithms tend to converge at a single optimum solutions. The convergence of the algorithm over a single optimal solution greatly depends upon the weights assigned to individual objectives and the solution may not be valid in case of change of weights.

ii) *Posteriori approach*: To overcome the above mentioned limitations of the priori approach, posteriori approach is preferred for multi-objective optimization problems. Posteriori approach consists of a sequence of steps to be followed to obtain the optimal solution as described below.

- Define objective functions.
- Apply suitable multi-objective optimization algorithm to obtain a dense spread of Pareto points.
- Select a single optimal solution from a set of Pareto points based on user requirement.

The specific characteristics of posteriori approach are

1) The multiple objective functions are not combined to form a single composite function and thus it is a multi-objective optimization technique in its true sense.
2) The algorithm does not converge over a single optimum solution but a dense spread of optimal solutions is obtained each of which corresponds to a particular order of importance of objectives.
3) In case of change of order of importance of objectives the user only requires to choose the most suitable optimum solution from Pareto optimal set of solutions. Thus this approach can very well suit the current manufacturing scenario which is extremely volatile.

The process of optimization using posteriori approach is presented graphically in Fig. 10.2.

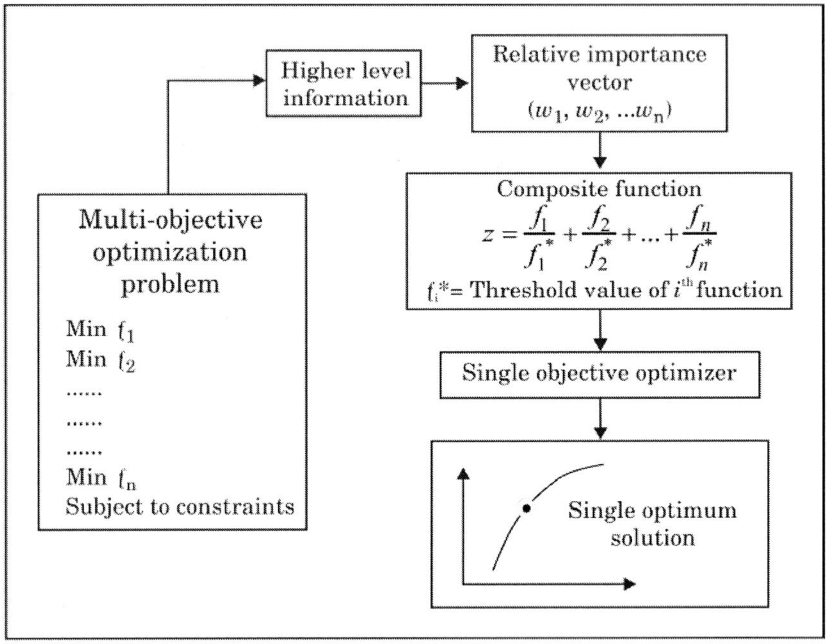

Fig. 10.1: The process of optimization using priori approach (Deb, 2011).

10.2. Methods of Formulating Combined Objective Function (*Z*) in Multi-Objective Optimization using Priori Approach

10.2.1. *Weighted sum method (Marler and Arora, 2004)*

The most common approach to formulate combined objective function (*Z*) in multi-objective optimization is the weighted sum method:

$$Z = \sum_{i=1}^{n} w_i F_i \tag{10.1}$$

It has particular application in group decision making and is used around the world in a wide variety of decision situations, in fields such as government, business, industry, healthcare, shipbuilding and education.

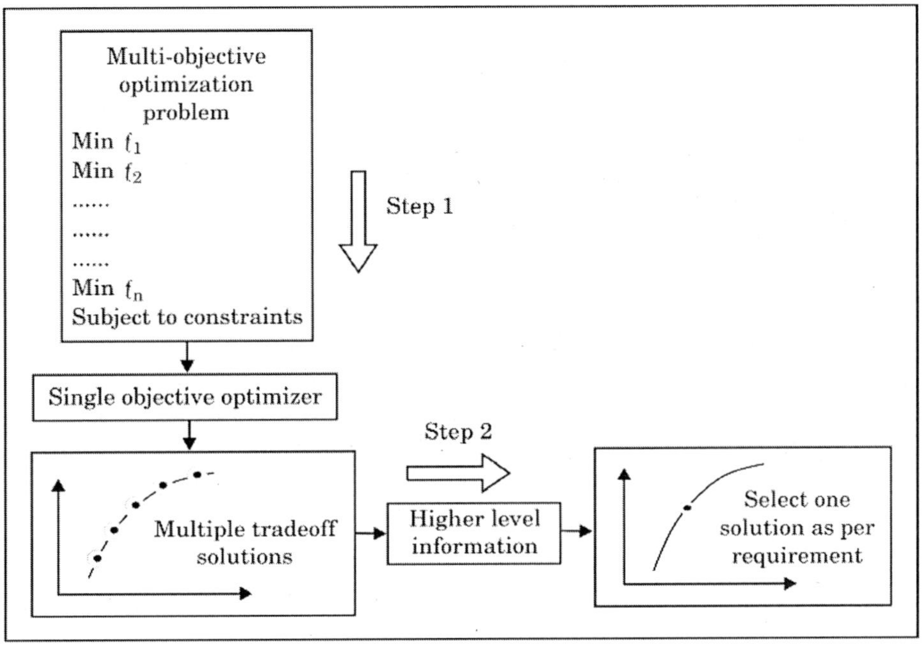

Fig. 10.2: The process of optimization using *posteriori* approach (Deb, 2011).

However, misinterpretation of the theoretical and practical meaning of the weights can make the process of intuitively selecting non-arbitrary weights an inefficient chore. To resolve this issue, researcher have developed systematic approaches to select weights. One of the popular methods for selecting the weights is Analytic Hierarchy Process (AHP). The analytic hierarchy process (AHP) is a structured technique for organizing and analyzing complex decisions, based on mathematics and psychology. The AHP (Saaty, 2008) is based on measurement through Pair-wise comparisons and is effective in dealing with complex decision making. AHP helps in complex decision making process and enables the decision maker to set priorities for arriving at the best possible decision. Pair-wise comparison is the most important stage in the AHP. A factor which is of significant importance for one person might not be of same importance for the other. To obtain Pair-wise comparisons, a relative measure is used. Every criterion is compared with other criteria and preference score is assigned to every paired comparison. The score indicates how many times the criteria under consideration is important or significant for the respondent over the other criteria to which it is compared with. Psychological experiments (Miller, 1955) have shown that individuals cannot simultaneously compare more than seven objects (with a margin of two). This is the main reason why Saaty (2008) has used nine as the upper limit for his scale. AHP also provides a consistency check to reduce bias in the decision making process. The AHP relies on knowledge and judgment of the experts/respondents to measure intangibles in relative terms.

The steps of AHP are

Step 1: *Pair-wise comparison*

To make comparisons, we need a scale of numbers that indicates how many times more important or dominant one element is over another element with respect to the criterion or property with respect to which they are compared. The respondent uses his knowledge and judgment to rate the factors against each other on a scale of nine. Initially the first alternative is compared with rest of the alternatives considering the first alternative as a reference point. Then the second alternative is compared with rest of the alternatives considering second alternative as a reference point and so on. Thus the AHP does not take a fixed reference point but treats all alternatives as reference points in order to minimize any bias which may be introduced through the selection of a single focus for the comparisons. The scale of relative importance introduced by Saaty (2008) is, 1 for equal importance, 3 for moderately more important, 5 for strongly more important, 7 for very strongly important, and 9 for absolute importance of one objective over another. The values 2, 4, 6, 8 are used to represent in between importance.

Step 2: *Preparing a Pair-wise comparison matrix*

The scores of all these comparisons are used to form a Pair-wise comparison (reciprocal) matrix. If the value for paired comparison of Criteria 1 and Criteria 2 (*i.e.,* r_{12}) is 'x', then the corresponding value for comparison of Criteria 2 with Criteria 1 (*i.e.,* r_{21}) will be '$1/x$'.

Step 3: *Determining weights*

Based on the Pair-wise comparison matrix, the AHP generates weights for every evaluation criteria. If M is the number of objectives, the geometric mean (*GM*) for every criterion (row wise) is calculated using following Equation.

$$GM_j = \left\{ \prod_{j=1}^{M} r_{ij} \right\}^{1/M} \tag{10.2}$$

After obtaining individual *GM*, summation of the individual *GM* is obtained.

The individual weights are then obtained by dividing individual *GM* with summation of *GM*. For instance, weight for the each objective w_j can be obtained as follows:

$$w_i = \frac{GM_j}{\sum GM_j} \tag{10.3}$$

Step 4: *Consistency check*

Consistency check must be carried out to ensure the consistency of the comparisons obtained through survey. The Pair-wise comparison matrix (say matrix A_1 prepared in step 2) is multiplied with the weight matrix (say matrix A_2, obtained in step 3) to obtain a new matrix (say matrix A_3).

Then every element of the matrix (A_3) is divided with the corresponding weight to obtain and a new matrix (say matrix A_4). The average of this new matrix (A_4) is the value of λ_{max}. Consistency index (*C.I*) is calculated using this value of λ_{max} with the following formula:

$$CI = \frac{\lambda_{max} - M}{M - 1} \tag{10.4}$$

Consistency ratio (*CR*) is the ratio of *CI* to the random consistency index (*RI*). The random index (*RI*) for up to 10 objectives is shown in Table 10.1.

If the *CR* is less than or equal to 0.1, the weights obtained using AHP are considered for further study otherwise the weights obtained are required to be modified by making suitable changes in the reciprocal matrix.

Table 10.1: Random index (RI) values

M	2	3	4	5	6	7	8	9	10
RI	0	0.90	0.58	1.12	1.24	1.32	1.41	1.45	1.49

Limitations of weighted sum method

1. Despite many methods for determining weights, a priori selection of weights does not necessarily guarantee that the final solution will be acceptable; one may have to resolve the problem with new weights. In fact, weights must be functions of the original objectives, not constants, in order for a weighted sum to mimic a preference function accurately.
2. It is impossible to obtain points on non-convex portions of the Pareto optimal set in the criterion space using weighed sum method.
3. Varying the weights consistently and continuously may not necessarily result in an even distribution of Pareto optimal points and an accurate, complete representation of the Pareto optimal set.

10.2.2. *Exponential weighted criterion*

In response to the inability of the weighted sum method to capture points on non-convex portions of the Pareto optimal surface, Athan and Papalambros (1996) proposed the exponential weighted criterion, as follows:

$$Z = \sum_{i=1}^{k} (e^{pw_i-1)} \cdot e^{pF_i(x)} \qquad (10.5)$$

where the argument of the summation represents an individual utility function for $F_i(x)$. Although large values of exponent '*p*' can lead to numerical overflow, minimizing Eq. (10.5) provides a necessary and sufficient condition for Pareto optimality.

10.2.3. *Weighted product method*

To allow functions with different orders of magnitude to have similar significance and to avoid having to transform objective functions, one may consider the formulation given by Eq. (10.6).

$$Z = \prod_{i=1}^{k} [F_i(x)]^{w_i} \tag{10.6}$$

10.3. Methods of Multi-Objective Optimization Using Posteriori Approach

Multi-objective optimization problems arise in a natural fashion in the engineering field. It should be preferable to optimize the objective functions all at once but, in general, they are in competition with each other and the optimization process has to search for the best optimal compromise solution. The primary goals in multi-objective optimization problem solution are: (a) To preserve non dominated points in the objective space and associated solution points in the decision space (b) To keep making algorithmic progress toward the Pareto front in the objective function space (c) To maintain diversity of points on the Pareto front and of Pareto optimal solutions (d) To provide the decision maker, the designer, with a large enough but limited number of Pareto points for selection.

Few popular methods of multi-objective optimization using posteriori are discussed below

10.3.1. *Non-dominated sorting genetic algorithm (Srinivas and Deb, 1995)*

Genetic algorithm (GA) possesses advantages that it does not require any gradient information and inherent parallelism in searching the design space, thus making it a robust adaptive optimization technique. Different versions of genetic algorithm including multi-objective genetic algorithm (MOGA), vector evaluated Genetic Algorithm (VEGA) and non-dominated sorting genetic algorithm (NSGA) can be applied to solve the multi-objective optimization problems. The flow chart of NSGA algorithm is presented below.

Steps of NSGA algorithm (Srinivas and Deb, 1995)

Step 1: *Parameter selection*

Selection of algorithm such as population size, string length = 16, Crossover probability = 0.8, Mutation probability etc. is crucial to govern the performance of NSGA algorithm.

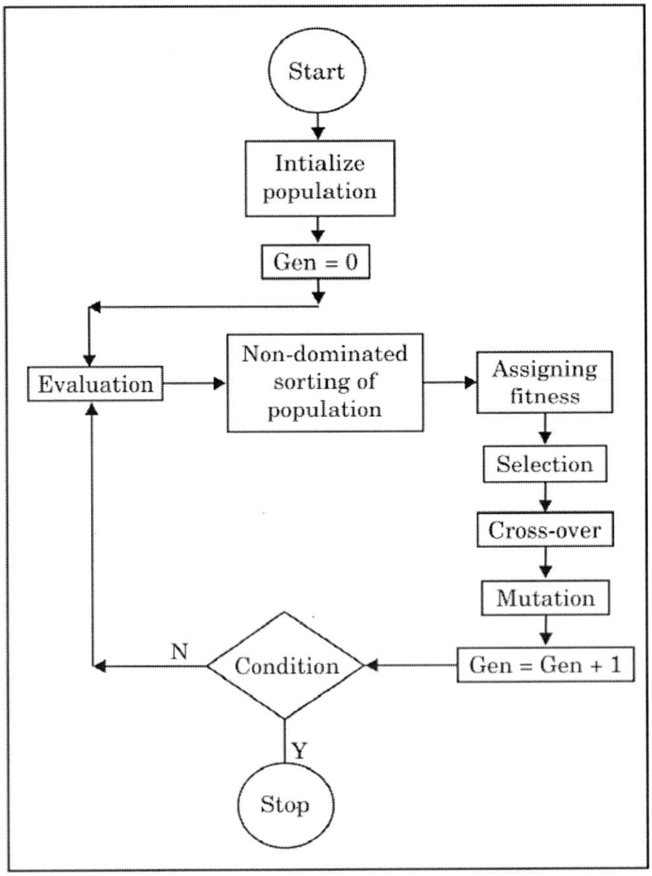

Fig. 10.3: Flowchart of NSGA algorithm (Srinivas and Deb, 1995).

Step 2: *Generate initial population of solutions and corresponding objective function values*

The values of all objective functions are evaluated for the randomly generated population of solutions.

Step 3: *Non-dominate sorting of solutions*

As shown in Fig. 10.4, in non-dominated sorting genetic algorithm, the crossover and mutation operators remain as like GA, but selection operator works differently from simple GA. In NSGA, a shared fitness is used for selection instead of actual fitness values of objective functions. The shared fitness value is calculated based on a non-dominated sorting and ranking procedure as demonstrated in Fig. 10.4 below.

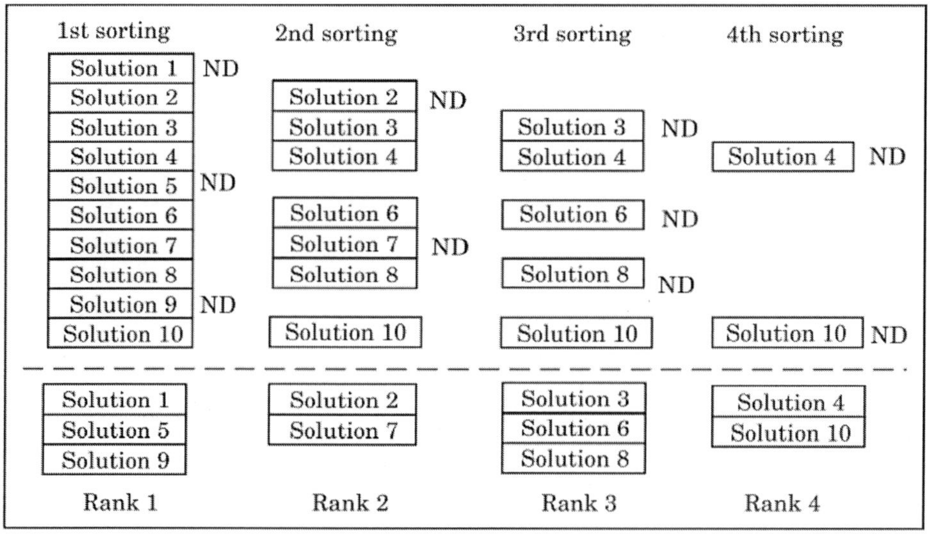

Fig. 10.4: Non-dominated sorting procedure (Srinivas and Deb, 1995)

As shown in Fig. 10.4, if we consider the case of two objectives, in first sorting, each solution is selected and checked whether it satisfies the rules given by Eq. (10.7) with respect to any other solution in the population

$$\text{Obj.1}[i] < \text{Obj.1}[j] \text{ and Obj.2}[i] < \text{obj.2}[j],\ i \neq j \qquad (10.7)$$

If the rules are satisfied for any one of the remaining solutions, then the selected solution is marked as dominated. Otherwise, the selected solution is marked as non-dominated. All the non-dominated solutions in the first sorting are ranked 1. Fig. 10.4 shows, solutions 1, 5 and 9 are non-dominated in first sorting, hence are rank 1 solutions. Eq. (1) is then again applied to remaining (dominated) solutions from the first sorting and non-dominated solutions identified in this second sorting are ranked 2. Solutions 2 and 7 are non-dominated in second sorting, hence are rank 2 solutions. The procedure is repeated until all the solutions are ranked. The subpopulation with rank 1, referred to as first front set, is assigned a dummy fitness value.

Step 4: *Determine normalized Euclidean distance of each solution*

Then, the normalized Euclidean distance of each solution is calculated with respect to all other solutions within this first front set using the formula:

$$d_{ij} = \sqrt{\sum \frac{\left(x_s^i - x_s^j\right)}{x_s^{max} - x_s^{min}}} \tag{10.8}$$

where, x_s the value of sth decision variable and i, j are solution numbers. x_s^{max} and x_s^{min} are upper and lower limits of the sth decision variable respectively.

Step 5: *Determine niche count of the solution*

A niche count (nc_i) provides an estimate of the extent of crowding near a solution and is calculated using the equation:

$$nc_i = \sum sh\left(d_{ij}\right) \tag{10.9}$$

Where, $Sh(d_{ij})$ is the sharing function values of all the first front solutions as given by equation:

$$Sh(d_{ij}) = \left\{1 - \left(\frac{d_{ij}}{\sigma_{share}}\right)^2\right\} \quad if \ d_{ij} < \sigma_{share} \tag{10.10}$$

$$= 0 \ \text{otherwise}$$

Where, σ_{share} is the maximum distance allowed between any two solutions to become members of a niche. The σ_{share} value in this equation is to be chosen appropriately.

Step 6: *Determine the shared fitness values of the solutions*

The shared fitness values (F) are then calculated by dividing the dummy fitness values by the niche count, which is given by Eq. (10.11).

$$Shared\ fitness\ (F_i) = Dummy\ fitness\ (f)\ /\ nc_i \tag{10.11}$$

After calculating the shared fitness values for the first front, a smaller value is subtracted from the minimum shared fitness value in this front and resultant value is given as the dummy fitness value (F_2) for all the next front (rank 2) solutions and the steps are repeated. This procedure is continued till the shared fitness values are calculated for all fronts.

Step 7: *Reproduction*

The process of natural selection is carried out in reproduction stage. Fit solutions are those having higher value of shared fitness. Applying the usual procedure of genetic algorithm for selection, fit solutions are copied for more number of times (proportional to their shared fitness values) and weak solutions will be eliminated.

Step 8: *Generate child solutions using crossover and mutation*

Like genetic algorithm, in NSGA also two strings are selected at a random from the mating pool. It is then decided whether to crossover using a parameter called crossover probability. If the crossover takes place then a random slicing point is chosen in the string. The sliced regions are then mixed to create two new strings. For each string element in each string in the mating pool the algorithm checks to see if it should perform mutation and if it should the string element is flipped (in case of binary strings). The mutation thus helps to prevent the population from stagnating and maintains the diversity throughout the iterations. To demonstrate the effectiveness of non-dominated sorting genetic algorithm, an application example of multi-objective optimization of grinding process is discussed in this section.

The optimization model for grinding process is formulated based on the analysis given by Wen *et al.* (1992). The four decision variables considered for this model are, wheel speed 'V_s' (m/min), work piece speed 'V_w' (m/min), depth of dressing '*doc*' (mm) and lead of dressing '*L*' (mm/rev).

The three objectives considered in this work are

a) Minimization of production cost 'C_T'($/pc).
b) Maximize the production rate in terms of work-piece removal parameter '*WRP*' (mm^3 / min N).
c) Minimization of surface roughness 'R_a' (mm).

However, keeping in view the specific requirement of finish grinding and rough grinding operation, these three objective functions are divided into two groups as follows:

For rough grinding operation following two objective functions are considered with the condition that the surface roughness value should not exceed 1.8 mm.

a) Minimization of production cost (C_T) in \$/piece
b) Maximize the production rate in terms of work-piece removal parameter '*WRP*' (mm^3/min N).

Whereas, for finish grinding operation following two objective functions are considered with the condition that the work-piece removal parameter should not be less than 20 mm^3/min N.

a) Minimization of production cost 'C_T'(\$/pc).
b) Minimization of surface roughness 'R_a' (mm).

These three objective functions 'C_T', '*WRP*', 'R_a' can be expressed in terms of the process variables as given by equations (10.12), (10.13) and (10.14) respectively.

$$C_T = \frac{M_c}{60p}\left(\frac{L_w + L_e}{V_w 1000}\right)\left(\frac{b_w + b_e}{f_b}\right)\left(\frac{a_w}{a_p} + S_p + \frac{a_w b_w L_w}{\pi D_e b_s a_p G}\right) + \frac{M_c}{60p}\left(\frac{S_d}{V_r} + t_1\right)$$

$$+ \frac{M_c t_{ch}}{60N_t} + \frac{M_c \pi b_s D_e}{60pN_d L V_s 1000} + C_s\left(\frac{a_w b_w L_w}{pG} + \frac{\pi(doc)b_s D_e}{pN_d}\right) + \frac{C_d}{pN_{td}} \quad (10.12)$$

Where, M_c is cost per hour labor and administration, L_w is length of work-piece, L_e is empty length of grinding, b_w is width of work-piece, b_e is empty width of grinding, f_b is cross feed rate, a_w is total thickness of cut, a_p is down feed of grinding, S_p is number of spark out grinding, D_e is diameter of wheel, b_s is width of wheel, G is grinding ratio, S_d is distance of wheel idling, p is number of work pieces loaded on the table, V_r is speed of wheel idling, t_1 is time of loading and unloading work pieces, t_{ch} is time of adjusting machine tool, N_t is batch size of the work pieces, N_d is total number of work pieces to be ground between two dressing, N_{td} is total number of work pieces to be ground during life of dressing, C_d is cost of dressing.

$$WRP = 94.4 \frac{(1 + (2\,doc\ /\ 3L))L^{11/19}(V_w / V_s)^{3/19} V_s}{D_e^{43/304}\, VOL^{0.47}\, d_g^{5/38}\, R_c^{27/19}} \tag{10.13}$$

Where, VOL = wheel bond percentage, d_g = grind size, R_c = work-piece hardness.

$$R_a = 0.4587\, T_{ave}^{0.30} \text{ for } 0 < T_{ave} < 0.254 \text{ else,}$$

$$R_a = 0.78667\, T_{ave}^{0.72} \text{ for } 0.254 < T_{ave} < 2.54 \tag{10.14}$$

$$T_{ave} = 12.5 \times 10^3 \frac{d_g^{16/27} a_p^{19/27}}{D_e^{8/27}} \left(1 + \frac{doc}{L}\right) L^{16/27} \left(\frac{V_w}{V_s}\right)^{16/27} \tag{10.15}$$

Following three constraints are considered in this optimization model

(a) Thermal damage constraint

The grinding process requires very high energy per unit volume of material removed. Whatever the energy that is concentrated within the grinding zone, it is converted into heat. The high thermal energy causes damage to the work piece, and it leads to the reduced production rate. The specific energy U is calculated by equation (10.16).

$$U = 13.8 + \frac{9.64 \times 10^{-4} V_s}{a_p V_w} + \left(6.9 \times 10^{-3} \frac{2102.4 V_w}{D_e V_s}\right) \times \left(A_0 + \frac{K_u V_s L_w a_w}{V_w D_e^{1/2} a_p^{1/2}}\right) \frac{V_s D_e^{1/2}}{V_w a_p^{1/2}} \tag{10.16}$$

The critical specific energy U^* at which burning starts is expressed in terms of the operating parameters as

$$U^* = 6.2 + 1.76 \left(\frac{D_e^{1/4}}{a_p^{3/4} V_w^{1/2}}\right) \tag{10.17}$$

The thermal damage constraint is then specified as

$$U^* - U \geq 0 \tag{10.18}$$

b) Wheel wear parameter constraint

Wheel wear parameter *WWP* (mm³/min. *N*) is related directly to the grinding conditions. For single-point diamond dressing, it is given by Eq. (10.19).

$$WWP = \left(\frac{k_p a_p d_g^{5/38} R_c^{27/29}}{D_c^{1.2/VOL-43/304} VOL^{0.38}} \right) \times \frac{\left(1+\left(doc/L\right)\right)L^{27/19}\left(V_s/V_w\right)^{3/19} V_w}{\left(1+\left(2doc/3L\right)\right)} \quad (10.19)$$

From equation (2) and (8) the wheel wear constraint is obtained as:

$$\frac{WRP}{WWP} - G \geq 0 \quad (10.20)$$

c) Machine tool stiffness constraint

Chatter results in poorer surface quality and lowers machining production rate. Chatter avoidance is therefore a significant constraint in selection of machining parameters. The relationship between grinding stiffness K_c (N/mm), wheel wear stiffness *Ks* (N/mm) and operating parameters during grinding is given below:

$$K_c = \frac{1000 V_w f_b}{WRP} \quad (10.21)$$

$$K_s = \frac{1000 V_s f_b}{WWP} \quad (10.22)$$

To avoid chatter during machining, the constraint given by equation (10.23) has to be fulfilled:

$$MSC - \frac{|R_{em}|}{K_m} \geq 0 \quad (10.23)$$

Where,

$$MSC = \frac{1}{2K_c}\left(1+\frac{V_w}{V_s G}\right) + \frac{1}{K_s} \quad (10.24)$$

Values of the constants and parameters used in the process parameter optimization of grinding process are given in Table 10.2. The variable bounds for all variables considered in this work are specified in Table 10.3. Considering the desired level of accuracy the bit length of four is suitable for all variables.

Table 10.2: Constants and parameters used in optimization of grinding process.

Notation	*Description*	*Unit*	*Value*
M_c	Cost per hour labor and administration	\$/hr	30
L_w	Length of work-piece	mm	300
L_e	Empty length of grinding	mm	150
b_w	Width of work-piece	mm	60
b_e	Empty width of grinding	mm	25
f_b	Cross feed rate	mm/pass	2
a_w	Total thickness of cut	mm	0.1
a_p	Down feed of grinding	mm/pass	0.0505
S_p	Number of spark out grinding		2
D_e	Diameter of wheel	mm	355
b_s	Width of wheel	mm	25
G	Grinding ratio		60
S_d	Distance of wheel idling	mm	100
p	Number of work-pieces loaded on the table		1
V_r	Speed of wheel idling	mm/min	254
t_1	Time of loading and unloading work-pieces	min	5
t_{ch}	Time of adjusting machine tool	min	30
N_t	Batch size of the work-pieces		12
N_d	Total number of work-pieces to be ground between two dressings		20
N_{td}	Total number of work-pieces to be ground during the life of dresser		2000
C_s	Cost of wheel per mm^3	\$	0.003
C_d	Cost of dressing	\$	25
VOL	Wheel bond percentage		6.99
d_g	Grind size	mm	0.3
R_c	Work-piece hardness	HRC	58
K_u	Wear constant	mm^{-1}	3.937×10^{-7}
R_{em}	Dynamic machine characteristics		1
K_m	Static machine stiffness	N/mm	100000
K_a	Constant dependent on coolant and grain type		0.0869

Table 10.3: Variable bounds of input variables.

Parameters	*Unit*	*lower limit*	*Upper limit*
V_s	mm/min	1000	2023
V_w	mm/min	10	22.7
doc	mm	0.01	0.137
L	mm/rev	0.01	0.137

Two examples are considered for optimization of grinding process problem as discussed below

Example 1

This example presents the multi-objective optimization of rough grinding process. The objectives being minimization of cost of production (C_T) and maximization of work piece removal rate (*WRP*) given by equations (10.12) and (10.13) respectively. The constraints are on thermal damage, wheel wear parameter, and machine tool stiffness as given by equations (10.17) (10.20) and (10.23) respectively. Also, for rough grinding operation the surface roughness value given by Eq. (10.14) should not exceed 1.8 μm. The steps of optimization using NSGA algorithm are discussed below.

Step 1: *Initialize the population*

In the example population size of 60 is considered. However, for demonstration purpose the values of only first 20 solutions are presented in this section. The objective function values for both objectives for the initial population are shown in Table 10.4.

Step 2: *Non-dominated sorting of solutions*

In this step the solutions in the population are sorted as dominated or non-dominated based on the condition specified by equation (10.7). To demonstrate the sorting procedure, initial solutions shown in Table 10.4 are considered. It is observed from Table 10.4, that out of 20 initial solutions, solution 1, solution 6, solution 7, and solution 17 are not dominated by any other solution in the population and hence are ranked as 1. Now, the sorting procedure will continue by eliminating the non-dominated solutions obtained in first sorting. The solutions identified as non-dominated solution in the second sorting are ranked as 2 such as solutions 2, 12, 13, 14, 18, and 20 in Table 10.4. This procedure will be continued until all 60 solutions are sorted and ranked. The ranking procedure emphasizes upon good solutions and increases the chance of selection of such solutions for the mating pool.

Table 10.4: Non-dominated sorting and ranking.

Sl. no.	*Solutions*	*C'*$_T$ *($/piece)*	*WRP'* *mm*^{3}*/min-N*	*Rank*
1.	1111110111011000001O	6.4737140	18.050175	1
2.	10001100001000000100	7.6449800	15.327719	2
3.	01000001111100100011	10.086645	15.245063	5
4.	00100010101100110011	9.5733250	13.144796	5
5.	10100111110100111100	9.4210280	15.553771	4
6.	11111010001101000100	8.6250980	25.323660	1
7.	11101101110101001011	6.2389700	17.189466	1
8.	00111111001110100011	11.820359	15.585501	5
9.	01110110110111110110	9.1461270	14.594395	5
10.	10000001100101100011	7.3714460	12.616397	3
11.	00010101010011011000	10.068557	7.5468130	6
12.	01101011101000010010	7.7460920	15.163872	2
13.	01111100110110101100	7.0886840	14.304452	2
14.	11011100000000010010	6.4069290	13.858267	2
15.	01001111110111000101	7.9264540	11.638927	4
16.	01111010111001010110	7.9663460	16.892221	3
17.	10101001100110111011	7.3886520	18.640226	1
18.	01010011000000101110	6.6813640	9.4310160	2
19.	01011110011101001000	9.7463740	15.888010	4
20.	10111100010100011001	7.4598780	17.052565	2

C'$_T$, *WRP'* = Cost of production and work-piece removal rate after assigning penalty.

Step 3: *Determination of shared fitness values*

The main objective to determine the shared fitness values of the solutions is to maintain the diversity of population. The shared fitness values are calculated using Eq. (10.11) and are shown in Table 10.5. As shown in Table 10.5, the solution 17 is having highest shared fitness value as it is rank 1 solution with relatively less value of niche count.

Step 4: *Reproduction*

In NSGA, the reproduction of the solution is based on the shared fitness value. The procedure for reproduction is explained in Table 10.6.

Table 10.5: Assigning fitness

Sl. no.	*F*	nc_i	f_i
1.	50.00000	7.970	6.260000
2.	5.876000	13.32	0.440000
3.	0.001289	9.590	0.000134
4.	0.001289	10.06	0.000128
5.	0.010500	6.390	0.001600
6.	50.00000	7.950	6.280000
7.	50.00000	8.100	6.170000
8	0.001289	9.320	0.000138
9.	0.001289	9.763	0.000132
10.	0.221000	7.24 0	0.030000
11.	0.001080	5.380	0.000022
12.	5.876000	13.74	0.420000
13.	5.876000	13.85	0.426000
14.	5.876000	13.78	0.426000
15.	0.010500	6.860	0.001530
16.	0.221000	7.160	0.030800
17.	50.00000	7.490	6.670000
18.	5.876000	13.43	0.437000
19.	0.010500	7.042	0.001400
20.	5.876000	13.91	0.422000

F = Dummy Fitness; nc_i = Niche count; f i = shared fitness

Table 10.6: Reproduction

Sl. no.	f_i	C_i	*R*	*Selected solutions*
1.	6.266000	0.0976	0.03	1
2.	0.440000	0.1044	0.1	2
3.	0.000134	0.104489	0.71	37
4.	0.000128	0.104491	0.91	51
5.	0.001600	0.1045	0.67	37
6.	6.280000	0.202	0.23	6
7.	6.170000	0.298	0.19	5
8.	0.000138	0.298	0.44	20
9.	0.000132	0.29856	0.74	37
10.	0.030000	0.299	0.4	16
11.	0.000022	0.299	0.9	51
12.	0.426000	0.305	0.03	0
13.	0.426000	0.31	0.34	16
14.	0.426000	0.3189	0.33	16
15.	0.001053	0.3189	0.69	37
16.	0.030800	0.423	0.72	37

Table 10.6: *Contd...*

Sl. no.	f_i	C_i	*R*	*Selected solutions*
17.	6.670000	0.43	0.16	5
18.	0.437000	0.43	0.22	6
19.	0.001400	0.436	0.23	6
20.	0.422000	0.4437	0.23	6

fi = Shared fitness; C_i = Cumulative probability; R = Random number

Step 5: *Updating the solutions*

The new solution (offspring) is obtained using crossover and mutation. In the present work, single-point crossover is used along with mutation in order to generate new population. Crossover probability is considered as 0.8 whereas mutation probability is considered as 0.2. The updated solutions after crossover and mutation at the end of first generation are presented in Table 10.7.

Table 10.7: New population after Crossover and Mutation

Sl. no.	*New population*	C_T *($/piece)*	*WRP* mm^3*/ min-N*
1	11101101110101001101	6.238133	17.706234
2	01101101010001000100	6.200752	7.724509
3	10101011111110100001	6.986998	29.444958
4	10101010001011000100	7.365420	20.428463
5	10111011111110100001	6.985902	30.408665
6	01111101110101001011	6.240968	13.701827
7	10111110001001001100	6.335504	19.752323
8	01111100110111101100	6.507544	15.724094
9	10111010001010000100	7.330585	19.859789
10	01111000001010000100	8.277761	16.560074
11	01111101000000000101	6.209629	7.257520
12	01111101001010000101	6.552129	17.928490
13	11011010001010000100	7.329537	21.107576
14	01111101000001000001	6.261650	7.088980
15	01111101110110001011	6.275218	14.542369
16	10111101000010010110	6.266798	15.983759
17	10111010001001001101	7.287586	18.666536
18	11101101000000000001	6.219595	7.041061
19	10111010001001001101	7.287586	18.666536
20	01111010001010000100	7.333235	17.318140

C_T = Cost of production; WRP = Work-piece removal rate.

Steps 1 to 5 are repeated till the convergence of the solution to obtain the Pareto-Optimal front in which all solutions are optimum and equally better. The Pareto-Optimal front is shown in Fig.10.5. Table 10.8 provides the optimal solutions obtained in Pareto-optimal set. This provides ready reference to the process planner to select the optimal combination of process parameters for rough grinding process as per the objective weights of his choice.

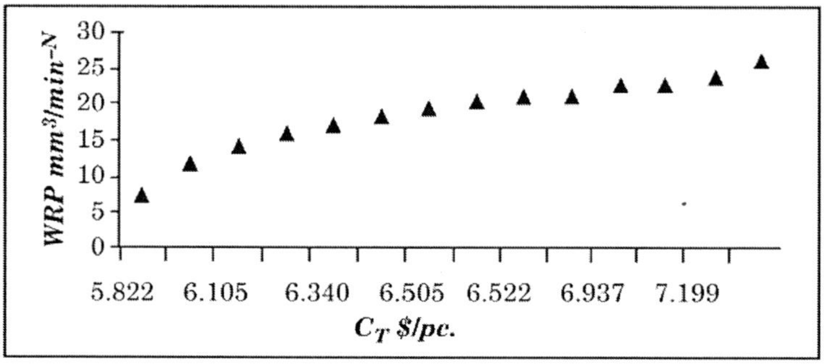

Fig. 10.5: Non-dominated front for rough grinding operation.

Table 10.8: Non dominated set for rough grinding operation (a_w = 0.1 mm, a_p = 0.050 mm/pass, $R_a \leq 1.8$ μm)

Sl. no.	*V_S m/min*	*V_W m/min*	*doc mm*	*L mm/rev*	*C_T $/pc.*	*WRP mm3/min-N*	*R_a μm*
1.	1759	22.70	0.014	0.014	5.822	7.54	1.102
2.	1759	22.70	0.030	0.014	5.890	11.05	1.533
3.	1825	21.00	0.034	0.067	6.010	15.31	1.670
4.	1891	18.60	0.055	0.043	6.350	16.47	1.730
5.	2023	18.60	0.055	0.018	6.350	17.25	1.760
6.	1957	16.96	0.047	0.084	6.500	18.21	1.660
7.	1957	16.96	0.051	0.084	6.520	18.64	1.700
8.	2023	16.96	0.051	0.084	6.520	19.17	1.680
9.	1957	16.96	0.059	0.084	6.560	19.51	1.770
10.	2023	13.69	0.084	0.084	7.200	21.97	1.790
11.	2023	10.82	0.096	0.137	7.990	24.79	1.780

V_S = Wheel speed; V_W = work piece speed; *doc* = depth of dressing; *L* = lead of dressing. R_a = surface roughness

Example 2

This example presents the multi-objective optimization of finish grinding process. The objectives being minimization of cost of production (C_T)

and minimization of Surface roughness (R_a) given by equation (10.12) and (10.14) respectively, satisfying thermal damage, wheel wear parameter and machine tool stiffness constraint given by Eqs. (10.17), (10.20) and (10.23) respectively, with work piece removal rate greater than 20 mm^3 / min-N. The steps of optimization using NSGA are followed in the same manner as discussed in the previous section.

The steps 1-5 discussed in example 1 are applied to obtain the Pareto-optimal front. Table 10.9 provides the optimal solutions obtained in Pareto-optimal set. The optimal combination of process parameters shown in Table 10.9 provides ready reference to the process planner for finish grinding process as per the objective weights of his choice. The Pareto-Optimal front is shown in Fig. 10.6.

Table 10.9: Non-dominated set for finishing grinding operation (a_w = 0.055 mm, a_p = 0.0105 mm/pass, $WRP \geq 20$ mm^3/min-N).

Sl. no.	V_S *m/min*	V_W *m /min*	*doc* *mm*	*L* *mm/rev*	C_T *$ /pc*	R_a *µm*	*WRP* *mm³/ min-N*
1.	2023	22.700	0.0100	0.1370	7.1100	0.790	20.010
2.	2023	18.190	0.0386	0.1165	8.0070	0.783	20.383
3.	2023	17.780	0.0645	0.1247	8.0700	0.775	20.490
4.	1990	15.735	0.0427	0.1160	8.6200	0.755	20.031
5.	2023	15.325	0.0386	0.1330	8.7200	0.753	20.939
6.	1957	14.500	0.0345	0.1370	8.9640	0.739	20.094
7.	1990	13.687	0.0468	0.1247	9.3030	0.736	20.470
8.	2023	13.277	0.0468	0.1330	9.4600	0.732	21.175
9.	2023	12.458	0.0427	0.1370	9.7890	0.706	20.860
10.	2023	11.229	0.0632	0.1042	10.488	0.696	20.360
11.	2023	10.819	0.0550	0.1247	10.688	0.684	20.701
12.	2023	10.000	0.0510	0.1330	11.199	0.659	20.590

V_S = Wheel speed; V_W = work piece speed; *doc* = depth of dressing; *L* = lead of dressing. *R*a = surface roughnes

10.3.2. *Non-dominated sorting genetic algorithm-II (NSGA-II)*

Although, NSGA has been successfully applied to many multi-objective optimization problems, the main criticisms of the NSGA approach have been as follows (Deb *et al.*, 2002):

1) ***High computational complexity of non-dominated sorting***: The currently used non-dominated sorting algorithm has a computational complexity of O ($M.N^3$) (where M is the number

of objectives and N is the population size). This makes NSGA computationally expensive for large population sizes. This large complexity arises because of the complexity involved in the non-dominated sorting procedure in every generation.

2) ***Lack of elitism*:** It has been proved that elitism can speed up the performance of the GA significantly, which also can help preventing the loss of good solutions once they are found.

3) ***Need for specifying the sharing parameter*:** Traditional mechanisms of ensuring diversity in a population so as to get a wide variety of equivalent solutions have relied mostly on the concept of sharing. The main problem with sharing is that it requires the specification of a sharing parameter (s_{share}). Though there has been some work on dynamic sizing of the sharing parameter, a parameter-less diversity-preservation mechanism is desirable.

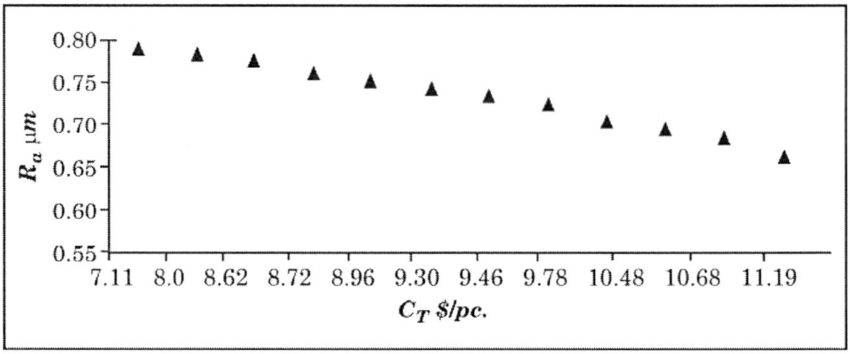

Fig. 10.6: Non-dominated front for finish grinding operation.

To overcome above mentioned issues, Deb *et al.* (2002) proposed improved version of NSGA known as NSGA-II. The characteristics of NSGA-II algorithm are:

- Encoding and decoding is eliminated as real coded parameters are used.
- Empirical relations instead of sharing parameter is used.
- Empirical relations for crossover and mutation instead of actual cross over of strings is used.
- Ensures elitism

The steps of NSGA-II are as follows

Step 1 – *Initialize population*: Initialize the population (N) based on the parameter range and constraint.

Step 2 – *Non-dominated sorting*: Sort the population by calculating rank of each individual based on non-domination criteria.

Step 3 – *Crowding distance*: *Calculate the crowding distance value front wise.*

Step 4 – *Tournament selection*: The individuals are selected using a tournament selection with crowded-comparison operator, *i.e.,* based on rank and crowding distance.

Step 5 – *Crossover and mutation*: Using simulated binary crossover (SBX) and polynomial mutation new parents and off-springs are found.

Step 6 – *Recombination and selection*: Offspring population (N) and current generation population (N) are combined ($N + N = 2N$) and the individuals of the next generation (N) from $2N$ are set again by selection using steps 2 to 4.

After initializing the population (N), the individuals in the populations are sorted based on non-domination into different fronts. The first front is completely non-dominant set in the current population and the second front is dominated by the individuals in the first front only and the front goes so on. Each individual in the each front are assigned rank (fitness) values based on front. Individuals in first front are given Rank-1 and individuals in second are assigned Rank-2 and so on. In addition to rank, crowding distance is calculated for each individual. The crowding distance is a measure of the spread between individuals, *i.e.,* how close an individual is to its neighbors. Large crowding distance will result in better diversity in the population. Parents are then selected from the population using tournament selection based on the rank and crowding distance. An individual having the rank lesser than the other or if crowding distance greater than the other one is selected.

Then from selected population using SBX, operator and polynomial mutation off-springs are generated. The population ($2N$) with the current population (N) and current off-springs (N) is sorted again based on non-domination and only the best N individuals are selected. The selection is based on the crowded comparison operator on the last front. The flow chart of the NSGA-II algorithm is shown in Fig. 10.7.

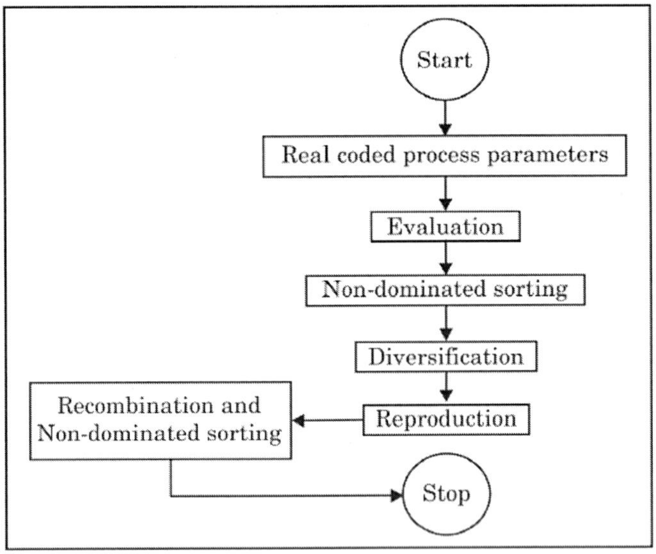

Fig. 10.7: Flowchart of NSGA-II (Deb *et al.,* 2002).

10.3.3. *Multi-objective Artificial Bee colony Algorithm*

As discussed in Chapter 4, the performance of artificial bee colony (ABC) algorithm in terms of convergence rate and accuracy of the solution is found superior over other non-traditional algorithms in many applications. ABC algorithm combines both, the stochastic selection scheme carried out by onlooker bees, and greedy selection scheme used by onlookers and employed bees to update the source position. Also the neighbor source production mechanism in ABC is similar to the mutation process, which is self-adapting. The random selection process carried out by the scout bees maintains diversity in the solution. The ABC algorithm is thus flexible, simple to use and robust optimization algorithm, which can be used effectively in the optimization of multimodal and multi-variable problems. To extend the application of ABC algorithm for multi-objective optimization with posteriori approach, concept of non-dominated sorting is incorporated in ABC algorithm as discussed through following steps (Pawar *et al.,* 2018):

Step 1: ***Parameter selection***

Algorithm specific parameters such as population size *i.e.,* number of food sources (equal to number of employed bees), number of onlooker bees, number of scout bees (5-30% of the colony size) are to be determined.

Step 2: ***Calculate the nectar amount for each food source***

The nectar amount represents the actual fitness value of the solution.

Step 3 to Step 6: These steps are same as those discussed in section 3 of this chapter.

Step 7: ***Determine the probabilities by using the shared fitness values evaluated in step 6:***

If the shared fitness of a food source "θ_i" is F_i, then the probability (P_i) of choosing this food source by an onlooker bee is expressed as:

$$P_i = \frac{\sum_{k=1}^{R} (1/f_k)^{-1}}{f_i} \tag{10.25}$$

Where, R is number of food sources.

Step 8: ***Calculate the number of onlooker bees, which will be sent to food sources***

The number of onlookers bees (N) sent to food source "θ_i" is calculated as:

$$N = P_i \times m. \tag{10.26}$$

Where, 'm' is the total number of onlooker bees.

Step 9: ***Determine new position of each onlooker bee:***

After watching the dances of employed bees, an onlooker bee goes to the region of food source "q_i" with the probability P_i. The position of the selected neighbor food source is obtained as below:

$$\theta_1 (c+1) = \theta_1(c) \pm \phi_1(c) \tag{10.27}$$

Where, "*c*" is number of generation. $\phi_i(c)$ is a randomly produced step to find a food source with a more nectar around 'θ_i'. Each food source is then updated if better position is obtained by onlooker assigned to that food source.

Step 10: ***Evaluate the best solution***

Position of the best onlooker bee is identified for each food source. The global best of the honeybee swarm in each generation is obtained and it may replace the global best at previous generation if it has better fitness value.

Step 11: ***Update the scout bee***

The worst employed bees, as many as the number of scout bees in the population, are respectively compared with the scout solutions. If the scout solution is better than employed solution, employed solution is replaced with scout solution. Else employed solution is transferred to the next generation without any change.

A flowchart of multi-objective artificial bee colony algorithm based on non-dominated sorting concept is shown in Fig. 10.8.

10.4. Improving the Quality Characteristics of Abrasive Water Jet Machining of Marble Material (Pawar *et al.*, 2018)

Abrasive water jet machining is used for making intricate decorative shapes or decorative profiles in marble material. Marble is a natural stone or rock resulting from metamorphism. Marble has number of applications in constructions, ceramics, paper, paint industries and decorative purposes. The diamond wire or saw cutters are conventionally used for cutting the marble material. During cutting the marble using diamond wire or saw cutter, the material grinds, rather than cut. Also various other drawbacks of conventional cutting of marble includes higher set up times, producing dust and noise, higher material wastage, unsuitability for profile cutting and cracking of material. Due to above mentioned issues encountered while conventional cutting of marble, attempts have been made for cutting of marble using non-traditional machining process such as ultrasonic

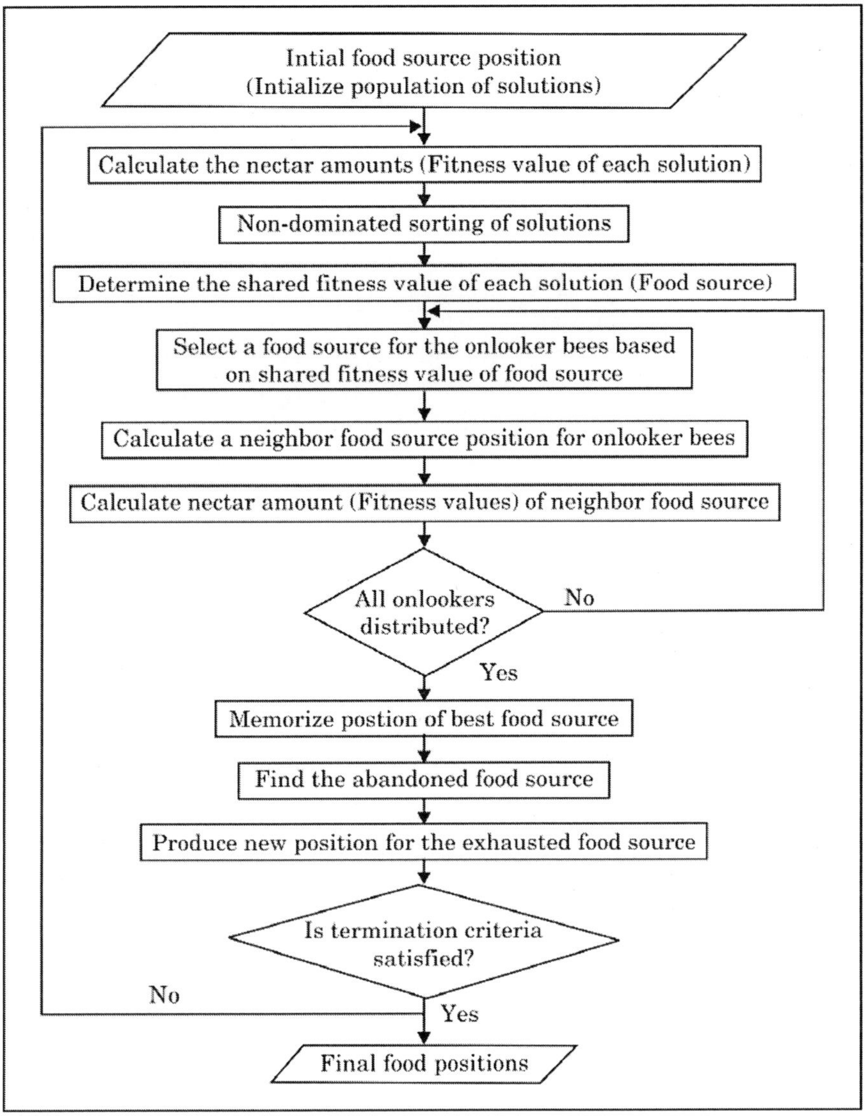

Fig. 10.8: Flowchart of Multi-objective ABC algorithm (Pawar *et al.*, 2018).

machining, abrasive water jet machining (AWJM), laser beam machining (LBM) etc. However, although ultrasonic machining can be applied to non-conductive as well as brittle materials, it is a slow and time consuming process with very high tool wear rate. Similarly, laser beam machining of marbles put constraint on the height of the work piece. Hence, AWJM has been claimed to be one of the most appropriate methods for cutting marbles due to its distinct advantages over the other cutting technologies, such as no thermal distortion of the work piece, high machining versatility to cut virtually any material, high flexibility to cut in any direction with small cutting forces, no risk of fire hazards no radiation emission, no tolerable noise levels etc.

In spite of several advantages AWJM offers while machining the marble material, achieving desired quality of cutting is a challenge. The performance measures of the quality of cutting includes kerf width (W), kerf taper angle (θ), and striations marks at bottom of machined surface (D). This case study provides the multi-objective optimization of above mentioned performance measures of AWJM process considering four process variables namely standoff distance (S), traverse speed (V), water jet pressure (P), and abrasive flow rate (A_f) for the work piece shown in Fig. 10.9.

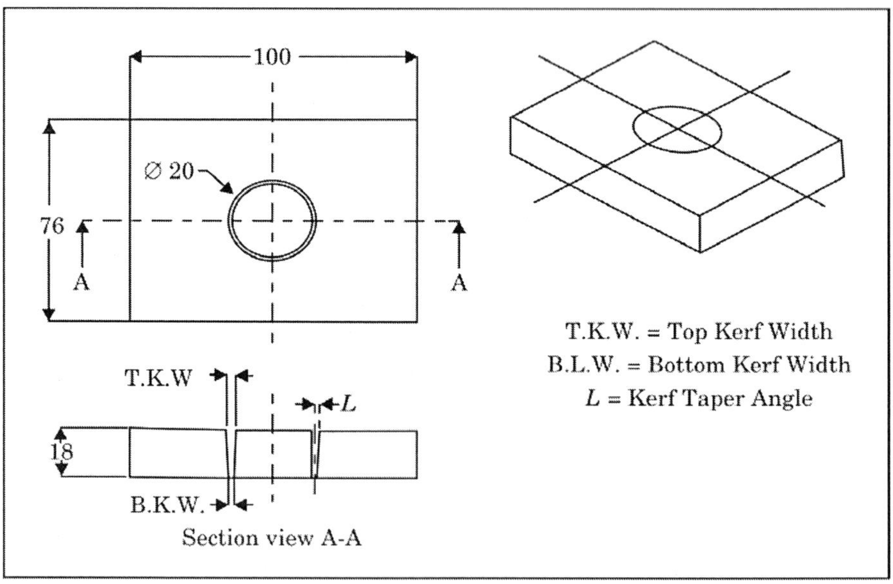

Fig. 10.9: Work-piece details and kerf geometry of an abrasive water jet cut

The ranges of process variables along with their coded values are shown in Table 10.10.

Table 10.10: Coded values of process parameters

Parameters	***Unit***	***Coded levels***				
		–2	***–1***	***0***	***1***	***2***
Stand off distance (S), (x_1)	mm	0.5	1	1.5	2	2.5
Traverse speed (V), (x_2)	mm/min	50	72.5	95	117.5	140
Water jet pressure (P), (x_3)	MPa	138	190	242	294	346
Abrasive flow rate (A_f), (x_4)	g/sec.	200	375	550	725	900

For work-piece in Fig. 10.9, the slug produced by AWJM is shown in Fig. 10.10. The slug clearly indicate the poor quality of cutting in terms of kerf width, kerf taper angle, and striations marks at bottom of machined surface (measured in terms of depth .of striation free surface.

Fig. 10.10: Slug produced by AWJM before optimization.

The mathematical models correlating the process variables (in coded form) with the three objective functions are given by Eq. (10.28) to Eq. (10.30)

$$\begin{aligned} W = {} & 1.1435125 + 0.00804167 x_1 - 0.017858 x_2 - 0.049429 x_3 \\ & - 0.003425 x_4 + 0.010625 x_1^2 - 0.02795 x_2^2 + 0.010275 x_3^2 - \\ & 0.0148625 x_4^2 + 0.02008125 x_1 \cdot x_2 - 0.0088375 x_1 \cdot x_3 - \\ & 0.04360625 x_1 \cdot x_4 + 0.03755 x_2 \cdot x_3 + 0.01001875 x_2 \cdot x_4 \\ & + 0.0251625 x_3 \cdot x_4 \end{aligned} \quad (10.28)$$

$$\theta = 0.9580575 + 0.0263904x_1 + 0.0732204x_2 + 0.467849x_3 - 0.258216x_4 0.0121178x_1^2 - 0.039688438x_2^2 + 0.03312406x_3^2 + 0.122430313x_4^2 + 0.22996188x_1 \cdot x_2 + 0.1483993x_1 \cdot x_3 + 0.057566875x_1 \cdot x_4 + 0.052045625x_2 \cdot x_3 - 0.156149375x_2 \cdot x_4 - 0.510349375x_3 \cdot x_4 \quad (10.29)$$

$$D = 6.925 + 1.2075x_1 + 1.10916667x_2 - 0.1975x_3 + 0.07x_4 + 0.265x_1^2 + 1.8025x_2^2 + 0.77x_3^2 + 0.24625x_4^2 + 0.08625x_1 \cdot x_2 + 1.33875x_1 \cdot x_3 + 0.78875x_1 \cdot x_4 + 2.22375x_2 \cdot x_3 + 0.67375x_2 \cdot x_4 + 2.57625x_3 \cdot x_4 \quad (10.30)$$

Now, following steps of multi-objective ABC algorithm are applied.

Step 1: ***Parameter selection***

Following algorithm specific parameters are considered in this work.

- Number of food sources: 20
- Number of onlooker bees: 50
- Number of scout bees: 1

Although 20 solutions are considered, for demonstration purpose only first five solutions are presented in next steps.

Step 2: ***Evaluate quality of each food source***

Quality of each food source is obtained by evaluating objective functions values as shown in Table 10.11.

Step 3: ***Non-dominated sorting of solutions***

Non-dominated sorting of all solution is done by applying criteria mentioned in Eq. (10.7). All solutions are ranked based on the level at which they attain non-dominated status. The ranking of solution is shown in Table 10.12.

Table 10.11: Nectar amount for each food source

Food source no.	*S (mm)*	*V (mm/min.)*	*P (MPa)*	A_f, *(g/sec.)*	*W (mm)*	θ^o	*D (mm)*
1	1.87	71.22	325.82	254.11	0.97	3.44	2.95
2	0.84	89.85	164.61	385.93	1.23	0.36	13.32
3	0.79	100.12	142.11	315.17	1.19	0.10	14.42
4	2.16	120.38	259.75	623.01	1.12	1.44	1.97
5	2.41	87.55	271.91	897.67	0.94	1.00	3.50

Table 10.12: Non-dominated sorting of solutions.

Solution no.	*(W) mm*	*(θ) °*	*(D) mm*	*Rank*
1	0.97	3.44	2.95	2
2	1.23	0.36	13.32	2
3	1.19	0.10	14.42	1
4	1.12	1.44	1.97	3
5	0.94	1.00	3.50	1

Step 4: *Determine normalized Euclidean distance of each solution*

The normalized Euclidean distance of each solution from every other solution is computed and is presented in Table 10.13.

Table 10.13: Normalized Euclidean distances

Sol. no.	*1*	*2*	*3*	*4*	*5*
1	0.000	0.971	1.088	0.835	1.009
2	0.971	0.000	0.189	0.934	1.189
3	1.088	0.189	0.000	1.017	1.326
4	0.835	0.934	1.017	0.000	0.553
5	1.009	1.189	1.326	0.553	0.000

Step 5: *Determine niche count of the solution*

Sharing function values shown in Table 10.14 are obtained by using Eq. (4). Niche count for solutions1 to 5 is obtained as 5.302, 7.822, 5.648, 9.898, and 7.570 respectively.

Step 6: *Determine the shared fitness values of the solutions*

Considering dummy fitness value for rank 1 solution as 50, the shared fitness value for solutions is computed. The dummy fitness value for second rank solutions is obtained to ensure that it should be less than rank 1 solution with minimum shared fitness value. In similar way the dummy fitness values of subsequent ranks 2, 3, and 4 are obtained as 3.586, 0.281, and 0.019 respectively. The shared fitness values for solutions 1 to 5 are 0.676, 0.458, 8.853, 0.028, and 6.605 respectively.

Table 10.14: Sharing function values

			Sharing function values		
Solutions	***1***	***2***	***3***	***4***	***5***
1	1.000	0.056	0.000	0.302	0.000
2	0.056	1.000	0.964	0.128	0.000
3	0.000	0.964	1.000	0.000	0.000
4	0.302	0.128	0.000	1.000	0.694
5	0.000	0.000	0.000	0.694	1.000

Step 7: ***Determine the probabilities based on the shared fitness values evaluated in step 6***

The probability (P_i) with which the onlooker bee is assigned to employed bee is obtained based on the shared fitness values. Values of P_i for solutions 1 to 5 are 0.012, 0.008, 0.157, 0.001, and 0.117 respectively.

Step 8: ***Calculate the number of onlooker bees, which will be sent to food sources***

The number of onlooker bees sent to an employed bee (N) obtained using Eq. (4.2) for solutions 1 to 5 are 1, 0, 8, 0, and 6 respectively.

Step 9: ***Determine new position of each onlooker bee***

Each employed bee is updated (for N times) to obtain positions of onlooker bees assigned to employed bees/food sources 1, 3, and 5. As mentioned in step 8, one onlooker bee is assigned to food source1, eight onlooker bees are assigned to food source 3 and six onlooker bees are assigned to food source 5. The positions of these onlooker bees are obtained as mentioned in Table 10.15.

The positions of onlooker bees for each food source are then compared with the position of the employed bee assigned to that food source. The comparison is made based on the combined objective function obtained considering equal weights to all objective functions. The combined objective function (to be minimized) is then evaluated as:

$$\min Z = \frac{W}{W^*} + \frac{\theta}{\theta^*} - \frac{D}{D^*} \tag{10.31}$$

Where, W^*, θ^*, D^* and are threshold values of kerf width, kerf taper angle, and depth of striation free surface respectively. The values of combined objective functions (Z) for employed and onlooker bees using Eq. (10.31) are shown in Table 10.14.

It is observed that combined objective function value (=101.14) of best onlooker bee for food source 1 is better than that of employed bee 1 (=114.87), and also combined objective function value of best onlooker bee for food source 5 (=23.02) is better than that of employed bee 5 (=33.71). Hence the positions of food source1 and 5 are updated. For food source 3, combined objective function value of employed bee (=3.58) is better than that of (=6.22) best onlooker bee position hence the position of the employed bee at food source 3 is retained. Table 10.16 shows the new set of first five solutions.

Table 10.15: Positions of onlooker bees.

Food source		*Positions of onlooker bee assigned to food source*	*(W) mm*	*(θ)°*	*(D) mm*	*Z*
1	1.	(1.89, 74.70, 322.14, 312.24)	1.00	3.02	2.66	101.14*
	1.	(0.77, 94.85, 158.32, 281.74)	1.18	0.18	14.79	6.22
	2.	(0.65, 98.07, 141.67, 622.35)	1.25	0.87	8.77	29.34
	3.	(0.81, 111.67, 144.14, 263.27)	1.09	0.21	13.47	7.25*
3	4.	(0.53, 99.59, 176.05, 325.33)	1.11	0.57	12.58	19.31
	5.	(0.74, 102.37, 161.56, 294.56)	1.13	0.22	13.48	7.56
	6.	(0.70, 101.41, 175.44, 352.96)	1.13	0.46	12.05	15.75
	7.	(1.04, 94.87, 142.11, 287.93)	1.25	0.30	15.84	10.15
	8.	(0.80, 101.75, 181.46, 559.45)	1.18	0.60	8.71	20.44
5	1.	(1.93, 80.77, 271.91, 895.20)	0.98	0.82	2.89	27.62
	2.	(2.35, 75.68, 274.64, 897.87)	0.90	0.84	3.05	28.39
	3.	(2.47, 106.42, 213.00, 469.85)	1.27	0.88	8.06	29.82
	4.	(2.43, 98.71, 215.44, 897.67)	0.98	1.40	8.38	47.09
	5.	(2.21, 83.34, 309.98, 897.67)	0.94	0.68	0.27	23.02*
	6.	(2.43, 91.73, 242.88, 900.00)	0.96	1.20	6.05	40.37

*Best onlooker bee position among onlooker bees assigned to particular food source.

Table 10.16: Set of updated food sources.

Food source no.	*S (mm)*	*V (mm/min.)*	*P (MPa)*	A_f, *(g/sec.)*	*W (mm)*	θ^o	*D (mm)*
1	1.89	74.70	322.14	312.24	1.00	3.02	2.66
2	0.84	89.85	164.61	385.93	1.23	0.36	13.32
3	0.79	100.12	142.11	315.17	1.19	0.10	14.42
4	2.16	120.38	259.75	623.01	1.12	1.44	1.97
5	2.21	83.34	309.98	897.67	0.94	0.68	0.27

This process maintains elitism as best among new solutions is compared with old solution and the one which is superior gets selected.

Step 10: *Evaluate the best solution*

The global best of the honeybee swarm is (0.79, 100.12, 142.11, 315.17) with W = 1.19 mm, q = 0.10° and D = 14.42 mm respectively.

Step 11: *Update the scout bee*

Since number of scout bee is one, a random solution is generated as (2.30, 66.90, 36.28, 64.16) with Z = 30.17. The worst among the updated solution is (1.89, 74.70, 322.14, 312.24) with Z = 101.14. As the randomly generated (scout) solution is better than the worst solution, the worst solution will be replaced by scout solution.

Step 12: *Obtain the set of Pareto-optimal solution*

The set of Pareto optimal solutions as shown in Table 10.17.

Although the parameters ranges available on machine are very wide, this study reveals that parameter values in some specific ranges of these parameters are only effective to achieve better process performance of wire electric discharge machining process with respect to the objectives considered. Value path plot shown in Fig. 10.12 provides the effective range of each parameter.

The Pareto front is shown in Fig. 10.11.

The application of Multi-objective ABC method to improve performance of AWJM process while machining of the work piece shown in Fig. 10.9, reveals that there is significant improvement in quality characteristics of product as shown in Fig. 10.13.

Table 10.17: Set of Pareto optimal solutions

Sr. no.	*(S) mm*	*(V) mm/ min.*	*(P) MPa*	*(A_f) g/ sec.*	*Kerf width (W) mm*	*Kerf taper angle (θ)°*	*Depth of striation free surface (D) mm*
1	2.50	62.02	189.57	670.38	1.145	0.005	11.794
2	2.12	84.00	181.58	490.62	1.293	0.001	13.024
3	1.71	50.00	344.76	900.00	0.799	0.062	0.000
4	2.15	75.79	198.31	448.73	1.276	0.001	12.121
5	0.61	140.00	240.21	578.05	0.958	0.007	0.329
6	2.34	77.38	163.82	614.45	1.274	0.003	14.542
7	0.99	84.07	157.17	285.70	1.255	0.015	15.991
8	2.02	50.00	223.98	208.49	1.186	0.045	12.370
9	0.96	91.99	153.43	303.37	1.229	0.005	15.171
10	0.68	77.79	138.00	281.43	1.299	0.001	17.028
11	0.77	140.00	200.06	900.00	0.985	0.004	0.000
12	2.50	50.00	343.75	900.00	0.623	0.079	0.000
13	2.48	50.00	186.29	722.95	1.059	0.014	10.785
14	2.50	50.00	344.76	900.00	0.622	0.075	0.000

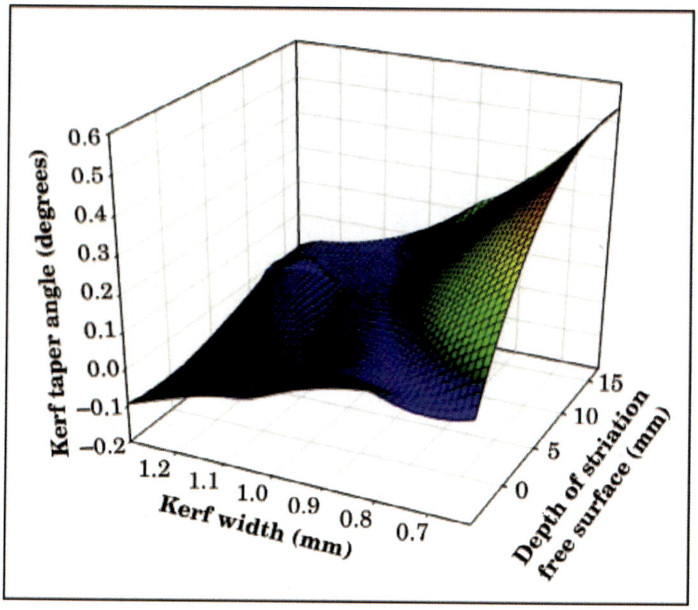

Fig. 10.11: Pareto front for three objective functions.

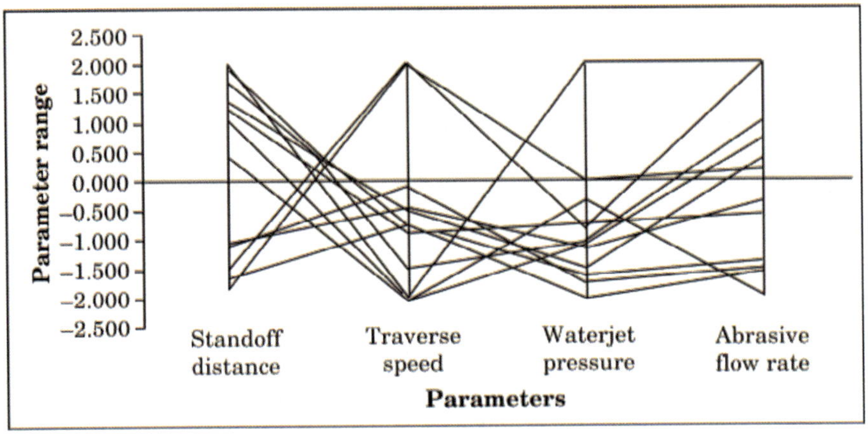

Fig. 10.12: Value path plot

Fig. 10.13: Slug for work-piece shown in Fig. 10.8 obtained by AWJM for solution no. 2 from the set of Pareto optimal solutions provided in Table 10.17.

References

Abazari A.M., Solimanpur M. and Sattari H. 2012. Optimum loading of machines in a flexible manufacturing system using a mixed-integer linear mathematical programming model and genetic algorithm. *Computers and Industrial Engineering* ***62(2):*** 469–478.

Acharya B.G., Jain V.K. and Batra J.L. 1986. Multiobjective optimization of ECM process. *Precision Engineering*, ***8*:** 88–96.

Alkin Y. and Erdal E. 2016. An Enhanced Artificial Bee Colony Algorithm with Solution Acceptance Rule and Probabilistic Multi-search. *Computational Intelligence and Neuroscience*. Article ID 8085953.

Arora P.K., Haleem A., Singh M.K. and Kumar H. 2013. Optimization of Cellular Manufacturing Systems Using Genetic Algorithm: *A Review, Advanced Materials Research*, ***622–623:*** 60–63.

Athan T.W. and Papalambros P. 1996. A note on weighted criteria methods for compromise solutions in multi-objective optimization. *Engineering Optimization*, **27**: 155–176.

Avraham M. and Gary D.K. 1989. An algorithm for computing the union, intersection or difference of two polygons. *Computer and Graphics*, ***13****(2)*: 167–183.

Bergh F. and Engelbrecht A.P. 2006. A study of particle swarm optimization particle trajectories. *Information Sciences*, ***176***: 937–971.

Bhangale P.P., Agrawal V.P. and Saha S.K. 2004. Attribute based specification, comparison and selection of a robot. *Mechanism and Machine Theory*, ***39***: 1345–1366.

Bhosale K.C. and Pawar P.J. 2018. Material flow optimisation of production planning and scheduling problem in flexible manufacturing system by real coded genetic algorithm (RCGA). *Flexible Services and Manufacturing Journal*, **31**(2): 381–423

Boumella N., Figueroa J.C. and Iqbal S. 2012. Enhancing Fuzzy Controllers Using Generalized Orthogonality Principle, *INTECH Open Access Publisher*.

Chakraborty P., Roy G.G., Das S., Jain D. and Abraham A. 2009. An improved harmony search algorithm with differential mutation operator. *Fundamenta Informaticae*, ***95***: 1–26.

Chen S.J. and Hwang C.L. 1992. Fuzzy Multiple Attribute decision-making, Methods and Applications. *Lecture Notes in Economics and Mathematical Systems, **375**.*

Dalvi A.M., Pawar P.J. and Singh T.P. 2016. Tool Path Planning, *In:* Hole Making Operations in ejector plate of injection moulds using modified shuffled frog leaping algorithm. *Journal of Computational Design and Engineering*, ***3***: 266–273.

Deb K. 2004. Optimization for Engineering Design: Algorithms and Examples, Prentice-Hall of India.

Deb K. 2011. Multi-objective Optimisation Using Evolutionary Algorithms: An Introduction. *In:* Wang L., Ng A. and Deb K. (*eds.,*) *Multi-objective Evolutionary Optimisation for Product Design and Manufacturing*, Springer, London.

Deb K., Pratap A., Agarwal S. and Meyarivan T. 2002. A Fast and Elitist Multi objective Genetic Algorithm: NSGA-II. *IEEE Transactions on Evolutionary Computation*, ***6****(2):* 182–197.

Dimopoulos C. and Zalzala A.M.S. 2000. Recent Developments in Evolutionary Computation for Manufacturing Optimization: Problems, Solutions, and Comparisons. *IEEE Transactions on Evolutionary Computation*, ***4****(2)*: 93–112.

Elbeltagi E., Tarek H. and Donald G. 2005. Comparison among five evolutionary based optimization algorithms. *Advanced Engineering Informatics*, ***19****(1)*: 43–53.

Eusuff M.M., Lanse K.E. and Pasha F. 2006. Shuffled frog-leaping algorithm: A memetic metaheuristic for discrete optimization. *Engineering Optimization*, ***38****(2):* 129–154.

Feng Z., Debao C. and Jiangtao W. 2016. An Improved Teaching-Learning-Based Optimization with the Social Character of PSO for Global Optimization. *Computational Intelligence and Neuroscience*, Article ID 4561507.

Gamila M.A. and Motavalli, S. 2003. A modeling technique for loading and scheduling problems in FMS. *Robotics and Computer-Integrated Manufacturing*, ***19****(1–2)*: 45–54.

Geem Z.W., Kim J.H. and Loganathan G.V. 2001. A new heuristic optimization algorithm: Harmony search'. *Simulation*, ***76****(2):* 60–68.

Ghaiebi H. and Solimanpur M. 2007. An ant algorithm for optimization of hole-making operations. *Computers and Industrial Engineering*, ***52****(2):* 308–319.

Goldberg D.E. 1989. Genetic Algorithms in search, optimization and machine learning, Addison-Wesley Longman Publishing Co., Inc. Boston, MA, USA.

Hasancebi O., Erdal F. and Saka M.P. 2010. An adaptive harmony search method for structural optimization. *Journal of Structural Engineering*, ***136***: 419–431.

Hashish M. 1989. A Model for Abrasive-Waterjet (AWJ) Machining. *Journal of Engineering Materials and Technology*, ***111***(2): 154–162.

Herrera F. and Lozano M. (2009). Fuzzy Evolutionary Algorithms and Genetic Fuzzy Systems: A Positive Collaboration between Evolutionary Algorithms and Fuzzy Systems. *In*: Mumford C.L. and Jain L.C. (*eds.*), *Computational Intelligence*, Springer-Verlag Berlin Heidelberg, pp. 83–130.

Holland J.H. 1975. Adaptation in neural and artificial systems, University of Michigan Press Ann Arbour MI.

Ingber L. and Rosen B. 1992. Genetic algorithms and very fast simulated re-annealing: A comparison. *Mathematical Computer Modelling*, ***16(11):*** 87–100.

Jain N.K. and Jain V.K. 2001. Modeling of material removal in mechanical type advanced machining processes: A state of the art review. *International Journal of Machine Tools and Manufacture*, ***41*:** 1573–1635.

Jain N.K., Jain V.K. and Deb K. 2007. Optimization of process parameters of mechanical type advanced machining processes using genetic algorithm. *International Journal of Machine Tools and Manufacture*, ***47(6)***: 900–919.

Jain P., Fenyes P. and Ritcher R. 1992. Optimal Blank Nesting Using Simulated Annealing. *Journal of Mechanical Design*, ***114(1)***: 160–165.

Kang Fei, Junjie Li and Sheng Liu 2013. Combined data with particle swarm optimization for structural damage detection. *Mathematical Problems in Engineering*.

Kennedy J. and Eberhart R. 1995. Particle Swarm Optimization. *Proceedings of IEEE International Conference on Neural Networks*, ***IV***: 1942–1948.

Kirkpatrick S., Gelatt C. and Vecchi M. 1983. Optimization by simulated annealing, *Science*, ***220(4598):*** 671–680.

Kolahan Farhad and Liang Ming 2000. Optimization of hole-making operations: A Tabu-search approach. *International Journal of Machine Tools and Manufacture*, ***40***: 1735–1753.

Koraboga D. and Basturk B. 2008. On the performance of artificial bee colony (ABC) algorithm. *Applied Soft Computing*, ***8(1):*** 687–697.

Kumar A., Singh B.K. and Patro B.D.K. 2016. Particle Swarm Optimization: A Study of Variants and Their Applications. *International Journal of Computer Applications*, ***135(5):*** 24–30.

Liang B., Zhen Z. and Jiang J. 2016. Modified shuffled frog leaping algorithm optimized control for air-breathing hypersonic flight vehicle. *International Journal of Advanced Robotic Systems*, ***13(6):*** 1–7.

Likert R. 1932. A Technique for the Measurement of Attitudes. *Archives of Psychology*, ***140***: 1–55.

Marco S., Paolo C., Daniela B., Riccardo C., Giancarlo M. and Gabriella P. 2018. Fuzzy Self-Tuning PSO: A settings-free algorithm for global optimization'. *Swarm and Evolutionary Computation*, ***39***: 70–85.

Marler R.T. and Arora J.S. 2004. Survey of multi-objective optimization methods for engineering'. *Structural and Multidisciplinary Optimization*, ***26***: 369–395.

Mehrdad M., Damangir E. and Mohammad F. 2007. An improved harmony search algorithm for solving optimization problems. *Applied Mathematics and Computation*, ***188(2):*** 1567–1579.

Metropolis N., Rosenbluth A., Rosenbluth M., Teller A. and Teller E. 1953. Equation of state calculations by fast computing machines. *Journal of Chemical Physics*, ***21(6):*** 1087–1092.

Miller G.A. 1955. The Magical Number Seven, Plus or Minus Two: Some Limits on our Capacity for Processing Information. *Psychological Review*, ***63***: 81–97.

Moscato P. 1989. On Evolution, Search, Optimization, Genetic Algorithms and Martial Arts: Towards Memetic Algorithms. *Caltech Concurrent Computation Program* (report 826).

Nguyen T.T. and Duong T.A. 2009. Comparing Three Improved Variants of Simulated Annealing for Optimizing Dorm Room Assignments. *International Conference on Computing and Communication Technologies, Vietnam*

Nighot M. and Pawar P.J. 2017. Experimental Investigation and Optimization of Control Parameters of Cooling System for an Industrial Robot Controller Cabinet', *International Conference on Manufacturing Excellence (ICMAX-2017).*

Parsopoulos K. and Vrahatis M. 2002. Recent approaches to global optimization problems through particle swarm optimization. *Natural Computing, **1***: 235–306.

Paulo J., Lashkari R.S. and Dutta S.P. 2002. Operation allocation and materials-handling system selection in a flexible manufacturing system: A sequential modelling approach. *International Journal of Production Research*, ***40****(1)*: 7–35.

Pawar P.J. 2010. Parameter optimization of some machining processes using non-traditional optimization techniques, *Ph. D. Thesis,* SVNIT, Surat.

Pawar P.J. and Deshmukh R.S. 2016. Multi-objective optimization of electroplating process using non-dominated sorting genetic algorithm. *Conference on Advances in Industrial Engineering, Manufacturing, Tool Design and Robotics,* **1**: 10–15.

Pawar P.J., Vidhate U.S. and Khalkar M.Y. 2018. Improving the quality characteristics of abrasive water jet machining of marble material using multi-objective artificial bee colony algorithm. *Journal of Computational Design and Engineering*, ***5****(3)*: 319–328.

Qin Q., Cheng S., Zhang Q., Li L. and Shi Y. 2015. Artificial Bee Colony Algorithm with Time-Varying Strategy. *Discrete Dynamics in Nature and Society*, Article ID 674595.

Rajurkar K.P., Zhu D., McGeough J.A., Kozak J. and De Silva A. 1999. New developments in electro-chemical machining. *Annals of CIRP*, ***48***: 567–579.

Rao R. and Yadava V. 2009. Multi-objective optimization of Nd: YAG laser cutting of thin super alloy sheet using grey relational analysis with entropy measurement. *Optics and Laser Technology*, **41**: 922–930.

Rao R.V. and Patel V. (2012). Multi-objective optimization of heat exchangers using a modified teaching-learning-based optimization algorithm. *Applied Mathematical Modelling*, ***37****(3):* 1147–1162.

Rao R.V., Pawar P.J. and Davim J.P. 2010. Optimization of process parameters of mechanical type advanced machining processes using a simulated annealing technique. *International Journal of Materials and Product Technology, **37**(1–2):* 83–101.

Rao R.V., Pawar P.J. and Shankar R. 2008. Multi-objective optimization of electro-chemical machining process parameters using a particle swarm optimization algorithm. *Journal of Engineering Manufacture, **222**(8):* 949–958.

References

Rao R.V., Savsani V.J. and Vakharia D.P. 2011. Teaching-learning-based optimization: A novel method for constrained mechanical design optimization problems. *Computer-Aided Design*, ***43***: 303–315.

Rao R.V., Savsani V.J., Vakharia D.P. 2012. Teaching-learning-based optimization: An optimization method for continuous non-linear large scale problems, *Information Sciences*, ***183****(1):* 1–15.

Ray T. and Liew K.M. 2003. Society and Civilization: An Optimization Algorithm Based on the Simulation of Social Behavior. *IEEE Transactions on Evolutionary Computation*, ***7****(4):* 389–396.

Rezaei Z. and Davoodi M. 2011. Multi-objective models for lot sizing with supplier selection. *International Journal of Production Economics*, ***130***: 77–86.

Saaty T.L. (2008). 'Decision making with the analytic hierarchy process'. *International Journal of Services Sciences,* ***1****(1):* 83–98.

Sadati N., Amraee T. and Ranjbar A.M. 2009. A global particle swarm-based simulated annealing optimisation technique for under-voltage load shedding problem. *Applied Soft Computing,* ***9***: 652–657.

Saka M. and Hasancebi O. 2009. Adaptive harmony search algorithm for design code optimization of steel structures'. *In: Geem Z. (ed.), Harmony search algorithms for structural design optimization. Springer-Verlag, Berlin, Heidelberg*, pp. 79–120.

Selvakumar A. and Immanuel T.K. 2009. Optimization using civilized swarm: Solution to economic dispatch with multiple minima . *Electric Power Systems Research*, ***79****(1):* 8–16.

Sharapov R.R. 2007. Genetic Algorithms: Basic Ideas, Variants and Analysis, Vision Systems: Segmentation and Pattern Recognition, pp. 407–422.

Sharma S., Sharma T.K., Pant M., Rajpurohit J. and Naruka B. 2015. Centroid Mutation Embedded Shuffled Frog-Leaping Algorithm. *Procedia Computer Science*, ***46***: 127–134.

Sharma T.K., Pant M. and Singh V. 2012. Improved local search in artificial bee colony using golden section search. arXiv preprint arXiv: 1210.6128; 2012.

Simpson T.W., Maier J.R.A. and Mistree F. 2001. Product platform design: Method and application. *Research in Engineering Design*, ***13****(1):* 2–22.

Srinivas N. and Deb K. 1995. Multi objective function optimization using non-dominated sorting genetic algorithms. *Evolutionary Computation*, ***2****(3):* 221–248.

Storn R. and Price K. 1997. Differential evolution - A simple and efficient heuristic for global optimization over continuous spaces. *Journal of Global Optimization*, ***11***: 341–359.

Sujono S. and Lashkari R.S. 2007. A multi-objective model of operation allocation and material handling system selection in FMS design. *International Journal of Production Economics*, ***105****(1)*: 116–133.

Szu H.H. and Hartley R.L. 1987. Non-convex optimization by fast simulated annealing'. *Proceedings of the IEEE,* ***75****(11):* 1538–1540.

Tizhoosh H.R. 2005. Opposition-Based Learning: A New Scheme for Machine Intelligence, *Int. Conf. on Computational Intelligence for Modeling Control and Automation-CIMCA' Vienna, Austria*, pp. 695–701.

Wang H., Wu Z., Rahnamayan S., Li C., Zeng S. and Jang D. 2011. Particle swarm optimisation with simple and efficient neighborhood search strategies. *International Journal of Innovative Computing and Applications*, ***3(2):*** 97–104.

Wang Hong-bo, Zhang Ke-peng and Tu Xu-yan 2015. A mnemonic shuffled frog leaping algorithm with cooperation and mutation. *Applied Intelligence,* ***43***: 32–38.

Wang Z.G., Rahman M., Wong Y.S. and Sun J. 2005. Optimization of multi-pass milling using parallel genetic algorithm and parallel genetic simulated annealing. *International Journal of Machine Tools and Manufacture,* ***45(15):*** 1726–34.

Wang X. 2009. Hybrid Nature-Inspired Computation Methods for Optimization, *Dissertation*, Helsinki University of Technology.

Wen X.M, Tay A.A.O. and Nee A.Y.C. 1992. Microcomputer based optimization of surface grinding process. *Journal of Material Processing Technology*, ***29***: 75–90.

www.mathworks.com

Yang Xin-She 2008. Nature-inspired Metaheuristic Algorithms, Luniver Press.

Yildiz A.R. 2013. A new hybrid artificial bee colony algorithm for robust optimal design and manufacturing'. *Applied Soft Computing,* ***13:*** 2906–2912.

Zadeh L.A. 1965. Fuzzy Sets, *Information and Control,* ***8***(3): 338–353.

Zhang C., Ni Z., Wu Z. and Gu L. 2009. Novel swarm model Quasi Oppositional Particle'. *International Forum on Information Technology and Applications*, pp. 325–330.

Zhang F., Zhang Y.F. and Nee A.Y.C. 1997 'Using Genetic Algorithms in Process Planning for Job Shop Machining. *IEEE Transactions on Evolutionary Computation,* ***4*****:** 278–289.

Zhao J. and Lv Li 2015. Two-Phases Learning Shuffled Frog Leaping Algorithm. *International Journal of Hybrid Information Technology*, ***8***(5): 195–206.

Zong-Sheng Wu, Wei-Ping Fu and Ru Xue 2015. Nonlinear Inertia Weighted Teaching-Learning-Based Optimization for Solving Global Optimization Problem, *Computational Intelligence and Neuroscience*, Article ID 292576,